Mycorrhizal Fungi - Utilization in Agriculture and Forestry

Edited by Ramalingam Radhakrishnan

Published in London, United Kingdom

IntechOpen

Supporting open minds since 2005

Mycorrhizal Fungi - Utilization in Agriculture and Forestry
http://dx.doi.org/10.5772/intechopen.91090
Edited by Ramalingam Radhakrishnan

Contributors
Baker Diwan Getheeth Aljawasim, Prashant Kaushik, Wawan Sulistiono, Taryono, Ozlem Altuntas, Srinivas Podeti, Dayakar Govindu, Anusha Duvva, Juan Francisco Aguirre-Medina, Jorge Cadena-Iñiguez, Juan Francisco Aguirre-Cadena, Kavita Chahal, Karima Bencherif, Terrafi Samia, Aranganathan Veeramani, Santhi Sudha Samuel, Opinder Singh Sandhu, Navjot Singh Brar, Vivek Kumar, Gurdeep Singh Malhi, Hari Kesh, Ishan Saini, Vaishali Gupta, Naveen Kumar Verma, Anand Chaurasia, Babita Rana

First published in London, United Kingdom, 2021 by IntechOpen
IntechOpen is the global imprint of INTECHOPEN LIMITED, registered in England and Wales, registration number: 11086078, 5 Princes Gate Court, London, SW7 2QJ, United Kingdom
Printed in Croatia

British Library Cataloguing-in-Publication Data
A catalogue record for this book is available from the British Library

Additional hard and PDF copies can be obtained from orders@intechopen.com

Mycorrhizal Fungi - Utilization in Agriculture and Forestry
Edited by Ramalingam Radhakrishnan
p. cm.
Print ISBN 978-1-83881-940-8
Online ISBN 978-1-83881-941-5
eBook (PDF) ISBN 978-1-83881-942-2

For EU product safety concerns:
IN TECH d.o.o., Prolaz Marije Krucifikse Kozulić 3, 51000 Rijeka, Croatia, info@intechopen.com or visit our website at intechopen.com.

Meet the editor

Dr. Ramalingam Radhakrishnan was born in India. He received several research awards and fellowships during doctoral and post-doctoral programs and made a significant contribution to the application of magnetic fields to the improvement of crop plants. His research was honored by the Chinese Academy of Science, which provided him financial support to present his findings at an international conference held in China. Professionally, he was employed as a post-doctoral researcher, research professor, and assistant professor at South Korean universities and research institutes. His major research finding is the utilization of microbes or elicitors to improve crops under environmental stress conditions and biological weed control. He has published several research and review papers as the main author in reputed journals, books, and conferences.

Contents

Preface

Several researchers have studied the association of mycorrhizae with several plant species and their significance in plant health. This book focuses on the production and applications of mycorrhizae in sustainable agriculture and forestry. The identification of soil microbes including fungi is essential to use those organisms for plant and human welfare.

In the first chapter, "Soil Metagenomics: Prospects and Challenges", the authors present a metagenomics study of soil microorganisms to identify categories of soil microorganisms. In the second chapter, "Arbuscular Mycorrhizal (AM) Fungi as a Tool for Sustainable Agricultural System", the authors discuss the application of arbuscular mycorrhizae in the soil to enhance uptake of nutrients, plant tolerance against drought, and soil quality. In the third chapter, "Advantages of Arbuscular Mycorrhizal Fungi (AMF) Production for the Profitability of Agriculture and Biofertilizer Industry", the authors explain the mitigation role of AM on adverse environmental conditions such as drought, heavy metal accumulation, salinity in the soil, extreme temperatures, and biotic stresses in plants. In the fourth chapter, "Production of Vegetable Crops by Using Arbuscular Mycorrhizae", the authors describe the use of mycorrhizae in soilless vegetables cultivation under a greenhouse system. In the fifth chapter, "The Role of Mycorrhizae on Seedlings and Early Growth of Sugarcane", the authors describe several varieties of sugarcane growth enhancement and mitigation of climate change by the colonization of mycorrhizae. In the sixth chapter, "Assessment of Biocontrol Potential of Arbuscular Mycorrhizal (*Glomus spp.*) against Damping-off Disease (*Rhizoctonia solani*) on Cucumber", the authors discuss the utilization of *Glomus* spp. to reduce *Rhizoctonia solani* pathogen infection and increase the growth of cucumber plants.

In the seventh chapter, "Native Arbuscular Mycorrhizal Fungi and Agro-Industries in Arid Lands: Productions, Applications, Strategies and Challenges", the authors discuss conventional and modern methods of AM fungal bio-fertilizers production including formulation and administration. The AM fungi influence the growth of several tree species in forests and improve soil quality. In the eight chapter, "Mycorrhizae Applications in Sustainable Forestry", the authors examine the significance of suitable AM fungi application for sustainable forest maintenance. In the final chapter, "Influence of Endomycorrhizal Fungi on the Growth of Tropical Plant Species", the authors discuss how fungi trigger the growth of tree species such as *Coffea canephora*, *Tabernaemontana donnell-smithii*, and *Chromolaena odorata*.

Dr. Ramalingam Radhakrishnan
Department of Botany,
Jamal Mohamed College (Autonomous),
Affiliated to Bharathidasan University,
Tiruchirappalli, Tamilnadu, India

Chapter 1

Soil Metagenomics: Prospects and Challenges

Prashant Kaushik, Opinder Singh Sandhu, Navjot Singh Brar, Vivek Kumar, Gurdeep Singh Malhi, Hari Kesh and Ishan Saini

Abstract

The better strategies to examine RNA or DNA from soil allow us to understand the microbial diversity and features in the soil, which are challenging to identify by typical culture techniques. In this direction, the literature on soil metagenomics and its usefulness is ever-increasing and so as its implementation experiences. Omics techniques are going to assist the metagenomics in achieving agricultural sustainability. In doing so, essential understanding on the reference soil would serve to help upcoming soil survey initiatives, lessening bias and raising objectivity. Although the interpretation of limited details has influenced microbial ecologists, the scope of methodological bias remains unfamiliar. A detailed catalog of functional genes and soil microorganisms does not yet exist for any soil. Overall, this chapter provides thoughts related to the soil metagenomics, its importance, and conventional methods of analysis, along with prospects and challenges of soil metagenomics.

Keywords: genomics, soil, microbes, metagenomics

1. Introduction

Soil is a robust and brilliantly vast ecosystem (2000–8.3 million bacterial species per gram). Therefore, it serves as a vast reservoir for microorganisms inhabiting in a niche that is different within the specific soil ecosystem, which can be pathogenic or beneficial [1–4]. Each proportion of soil whether in grasslands, forests, or deserts (i.e., sand, silt, clay, and organic matter) offers habitats for nematodes and a large number of microbes that vary from bacteria and are also useful in nutrient cycling [5–8]. Moreover, the distinct microhabitat dwelled by microorganisms with the capability to adjust and established their colony to the specific niche [9]. The crucial factors which influence the microbial load in the soil ecosystem include soil pH, organic compound, and temperature [10–12]. The chemical or physical activity does not merely determine the development of soil but the constant unfolding of different microbial species, which include or may improve the attributes of soils, regarding the development of function and structure [13, 14]. Soil supplies protection to different soil harboring animals, reptiles, and insects, along with a tremendous number of microbes inside the soil aggregates [15].

In this direction, the field of metagenomics continues to be a ground-breaking technology, which has made it possible to explore microbial diversity with its full potential [16]. Besides the soil ecosystem, microbes could quickly react to anthropogenic pressures, making it feasible to be an indication of soil quality as well as wellness [17, 18]. Lately, efforts have been attempted to determine genes from environmental samples via culture-independent techniques [19, 20]. However, they had been amplified or perhaps recognized due to their similarity to the earlier identified genes, that invalidates for exploiting novel elements of metal resistance [21, 22]. As the development of culture-independent metagenomic methods, it has been employed to evaluate the soil microbial community as well as enhances our awareness of the soil ecosystem [17, 20]. Furthermore, the soil microbial communities are primarily made up of some dominant species and numerous other rare taxa [23].

The ones with a low abundance might be from some novel microbial lineages and might play a vital function in biogeochemical interactions of the soil–plant system [24]. Therefore, the information obtained from the full metagenomic sequencing is crucial to expose the genomic data of low abundance populations as well as to disclose their activity in the soil [6]. It has been effectively released into investigating numerous varied microbial niches in the human gut, grassland soil, and aquatic ecosystem [25–27]. Furthermore, attempts have been established to evaluate the abundance of soil microbes as well as the genes involved in heavy metals' opposition from agricultural soils [28, 29]. Additionally, soil metagenomics beyond estimating the soil microorganisms can also help in getting a concept about the soil and its habitat based on the different soil types [30]. This chapter provides the importance of soil metagenomics and standard methods of analysis, along with challenges and prospects of soil metagenomics.

2. Soil health and metagenomics

Soil is an interconnected system because of its microorganisms despite getting incredible and unique capability to adjust to life changes; soil microbes are hypersensitive to land management and also weather changes [1, 11, 14]. Based on this information, our ancestors learned the ability to grow plants and created different cultivation methods like inoculating mycorrhizal fungi with food and floral crops to decrease the impact of soil-borne diseases [31–34]. With a most varied ecosystem with a composition of known and unknown microbial species, the soil provides an ecological niche [34, 35]. The biochemistry of soil reflects many anonymous functions that are a lot essential for sustenance of life [35–37]. Nevertheless, the latest technologies utilizing heavy machinery and management methods intensified agriculture and have resulted in the degeneration of the cultivable farmlands through damage of fertility, soil structure as well as the soil microbial life [38–40]. In a nutshell, lots of arable areas have switched to uncultivable or saline soil [38]. Agricultural land is simultaneously getting forfeited to nonagricultural uses [39]. The generation of soil, which primarily contains carbon twice as much as the atmosphere, is a complex phenomenon and requires lots of years for the formation of 1 cm topsoil [41]. Metagenomics data can be used to investigate the gene sequencing helpful in microbial symbiosis, as this is the most ancient symbiosis of nature of around 400 million years [42]. With the increased population pressure, the concern around worldwide sustainability also increases. Therefore, improving and sustaining the qualities of soil is an utmost concern for many years. Thus, soil health gets among the most crucial aspects of agriculture [8, 24].

Metagenomics offers an entirely new method of looking at the microbial community that has transformed contemporary microbiology and also has the potential to revolutionize comprehension approaches of the various ecosystems [43, 44]. In metagenomics, the strength of genomic examination is put on to whole populations of microorganisms [45]. Metagenomics approaches are throwing light on the myriad abilities of microbial communities that operate the planet's energy and nutrient cycles and form the evolution of life [46, 47]. Metagenomics is anticipated to produce awareness of microbial interactions; therefore, it is used to enhance human well-being, energy production, and food security [48]. Metagenomics combines the strength of genomics, systems biology and bioinformatics and power of genomics within the research of communities generates an unparalleled ability [43, 45]. Metagenomics, still a very new science, but has produced insightful information about the microbial community due to its radically unique means of realizing the microbial world [49, 50]. The diverse test of DNA may subsequently be analyzed directly, or even cloned into a type maintainable in lab bacteria, developing a library which has the genomes of all of the microbes present in that environment [51–53]. Nevertheless, the launch of the culture-independent approaches eliminates the obstacles and barriers in understanding the environmental samples [30].

Metagenomics initially targeted the shotgun sequencing; these days it's just as helpful for the scientific studies regarding marker genes *viz.* 16S rRNA by employing NGS (next-generation sequencing) systems, by extracting the specific region of DNA encoding 16S rRNA which is then amplified, sequenced, and identified based on similarities in gene sequence available in public databases [54, 55]. NGS, along with polymerase chain reaction (PCR), and DNA fingerprinting techniques have become increasingly rapid, effective, sensitive, and cost-efficient [55]. Culture-independent tactics are needed on the immediate extraction of soil DNA and later check out the genes encoding rRNA [56]. The exploration of following generation sequencing as well as analysis has accomplished in revealing the undiscovered microbial framework in a variety of earth ecosystems [57–59]. A comprehensive research of the soil metagenome provided the useful characterization of soil microorganisms linked to the genes in nutrient cycling [58]. Nevertheless, efforts are now being directed in exploring the predictions of gene operates in conditions of the actual role of theirs in situ, particularly in the soil, where metagenomes can easily be caught within biofilms [60, 61].

3. The need for microbial identification and characterization

It is well recognized that the microorganisms have an abundant quantity and diversity than other organisms on the planet [4]. Nevertheless, the division of the microbial diversity at global scales is still partially understood. The microbial diversity and composition structure are significantly affected by environmental elements [62]. As a result, indexing, cataloging and proof of the microorganisms are prerequisites for the exploration [62]. Microbial diversity in any habitat is more related to the substantial amount of species existed at a specific time [63, 64]. As the earth microbial community plays essential roles in soil health management, agro-ecosystem, accessibility of growing nutrition as well as turnover tasks of organic material in soil, they are hugely influenced by both anthropogenic and natural activities [65, 66]. For instance, many microbes that are helpful to the ecosystem services are currently threatened because of inferior agricultural practices, local weather transforming patterns, ground as well as land degradation, etc. [67, 68]. In recent years, the use of artificial fertilizers, herbicides, fungicides, and other pesticides has resulted in the deterioration of the soil microflora and diversity [7, 31, 69]. Therefore, the microorganisms with the changing atmosphere will offer a broader picture of the way the microbes are shifting the functional characteristics of soils and their flourishing in the endangered ecosystems [69, 70].

4. Metagenomics for sustainable agricultural practices

Nowadays, most of the environmental focus in agriculture is on achieving agricultural sustainability. Many metagenomic initiatives have been completed in the area of agriculture but do not hold some promise to assist the marginal farmers [71–73]. Therefore, productive scientific studies are required, which might be used the growers' income and help agriculture [74]. The latest advances in the soil metagenomics emerge as an extraordinary area of research because of the assignment of understanding the associated microorganisms in development and plant growth [75, 76]. Likewise, restoration of the microbial population was determined to improve grain yield as well as soil health [77]. Metagenomics can predict the soil microbes' structure and the impact on microbial groups of connected niches [9, 35].

Sustainable agricultural methods consist of different microhabitats with excellent environmental fluctuations and genetic biodiversity [78]. Reports from agrarian soils confirmed that there are high microbial stock and plant development promotion pursuits [79]. Many studies are showing the latest metagenomic improvements in agriculture [76–78]. Soil microbes play a crucial part for triggering the plant development, stress reactions, as well as defense in vegetation [80, 81]. Understanding the connection between the soil microbiota and plants using soil metagenomics is hugely advantageous in developing the crop systems [1, 82]. Metagenomics research of the soils supplemented with organic manures from several farm animals will be a lot valuable in formulating the fertilization tactics [12, 65]. For renewable agricultural production, helpful microbes of agricultural value can function as an essential alternative [39, 73]. Metagenomics compensation can address basic restorative questions associated with agriculturally significant microorganisms [83].

Direct DNA extraction and characterization through PCR and metagenomic survey have developed the study of soil ecosystem [81]. Applying metagenomics of plant-microbial association can be used to study interaction with beneficial microbes among pathogenic strains, infect recently, profitable endosymbionts inside these beneficial microbes (AMF) like nitrifying bacteria, (phosphate solubilizing bacteria) PSB and plant growth-promoting abilities are found [84]. The specificity of plant-microbial symbiosis development can be easily understood at the molecular level both for agronomic and horticulture crops, using forward and reverse genetic approaches [85]. As reported earlier, microbial inoculation has the potential of increasing plant production and sustainability in agricultural fields, so the metagenomics study can reveal the distinct microbial strains interacting with which chemical compound in the mycorrhizospheric soil and to acknowledge community structure, horizontal gene transfer analysis and phylogeny of microbes interacting with other environmental factors [86, 87]. Also, the exact niche information of microbial communities infecting soil adhering to the roots, surface between roots and soil, the surface of roots, or colonizing with the roots, can be drawn [87]. As we know this interaction is bi-directional, plant gets the essential nutrients from the soil, and in return, these rhizospheric and/or root microbiotas get the photosynthesis-derived organic compounds, and this process is known as rhizodeposition [88]. Thus, this crucial symbiosis that underlying plant-microbe community associations can be easily implicit by metagenomics for agricultural purposes because the NGS which can determine the relative abundance of microbe whether it is culturable in laboratory conditions or not [89]. In addition, the results of hundreds of samples simultaneously can be obtained on the same day as the samples are loaded [89].

5. Amplicon sequencing and bioprospecting of metagenomes

In the amplicon sequencing study, first, soil DNA is extracted, and next, 16S/18S rDNA sequences are amplified using a specific set of particular primers targeting variable areas of 16S/18S rDNA, accompanied by filtration of fragments using magnetic beads [54, 90–92]. Consequently, adapters are ligated, and the library of fragments (clones) is amplified along with the samples are sequenced utilizing NGS platform (**Figure 1**). The dataset obtained after sequencing is used for the identification of microbial diversity [54, 55]. Using NGS and the related software, it is doable to solve extremely complicated microbiota compositions with greater precision and to relate the microbial ecosystem of the soil [16, 55, 92], although, it must be considered for accurate data and analysis interpretation while choosing amplicon sequencing working with marker genes [92].

From the start of metagenomics, the study of novel metabolite/biomolecule (DNA polymerases, cellulases, lipases/esterases, chitinases and antibiotics) from the microbial assembly was its first application, and this has advanced with the development of NGS techniques for calculating comparison between community metagenome, meta-proteomics, and meta-transcriptomics [93, 94]. Techniques for recovering novel metabolite that comprise cloning of the microbial DNA from the environment then constructing a small/large-insert libraries, which can be done either by function-based or sequence-based screening of metagenomic libraries are shown in **Table 1**. The resulting metagenomic libraries subsequently transformed in several hosts like Escherichia coli (mostly), *Sulfolobus solfataricus*, *Thermus thermophilus*, *Streptomyces*, and *Proteobacteria* show significant differences in expression modes [108–111].

6. Soil metagenomics: pipelines and outputs

Metagenomics was initiated with the aim of DNA cloning and screening, and now it has made significant advances in microbiology, evolution, and ecology [91, 92]. These first projects not merely proved the concept of the metagenomics but additionally unraveled enormous gene diversity within the microbial world. The various steps in soil metagenomics are enlisted below and shown in **Figure 2**.

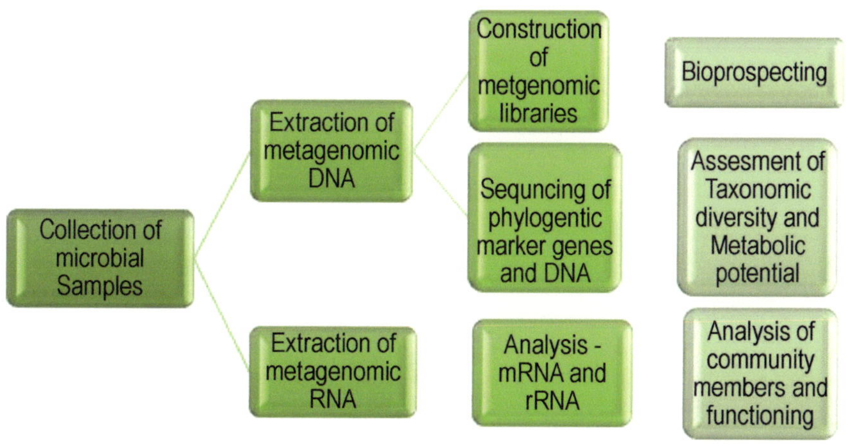

Figure 1.
Metagenomic analysis of environmental microbial sampling based on nucleic acids.

Origin	Sequencing platform/ amplicon analysis technique	Total sequencing size	Country	Results	References
Potato field	Pyrosequencing	1674 OTUs	USA	Identification of potato soil-borne pathogens	[95]
Soil of 3 islands in the Yellow Sea	Pyrosequencing	10,166 reads	South Korea	Wood decomposing, plant-parasitic, endophytic, ectomycorrhizal and saprotrophic fungi	[96]
6 sites of forest and grassland soils	Pyrosequencing	598,962 sequences	Germany	Identification of 17 bacterial phyla and 4 proteobacterial classes	[97]
Pea field	Pyrosequencing	55,460 sequences	Denmark	Fungal species, diversity, community composition of phylum Ascomycota, and Basidiomycota	[98]
Hitchiti *Pinus* forest, prior used as cotton cultivation	Deep Ion Torrent sequencing	>3,000,000 sequences	USA	12 fungal strains identification	[99]
Sossego copper mine	Pyrosequencing	10,978 OTUs	Brazil	36 bacterial phyla and five proteobacteria classes	[100]
Riverine Wetland soil	Illumina system	1872 OTUs	USA	56 different bacterial phyla	[101]
Solid biomedical dumpsites	Illumina system	1,706,442 sequences	Tanzania	31 bacterial phyla belonging to aromatic hydrocarbons degraders, chitin degraders, chlorophenol degraders and atrazine metabolizers	[102]
Grave-soil human cadavers	Illumina system	1,729,482 reads	USA	45 decomposing microbes identification	[103]
Zea mays fields	Illumina MiSeq	2,453,023 reads	UK, France, Italy	Comparative account of soil microorganisms of three different sites	[104]

Origin	Sequencing platform/ amplicon analysis technique	Total sequencing size	Country	Results	References
Pepper field	Illumina system	4147 OTUs	Spain	Studying soil-borne pathogens	[105]
Solid waste dumping site, Chite river site, Turial river site, Tuikual river site	Illumina system	111,3884 sequences	India	Identified 27 proteobacteria and bacteroidetes	[106]
Tomato	Illumina amplicon sequencing analysis and phytohormone measurements	337,961 high-quality reads and 647 fungal OTUs	Denmark	Identification of 27 endophytic fungi and root hormone quantification	[107]

Table 1.
Examples of soil amplicon sequencing done so far covering different habitat types.

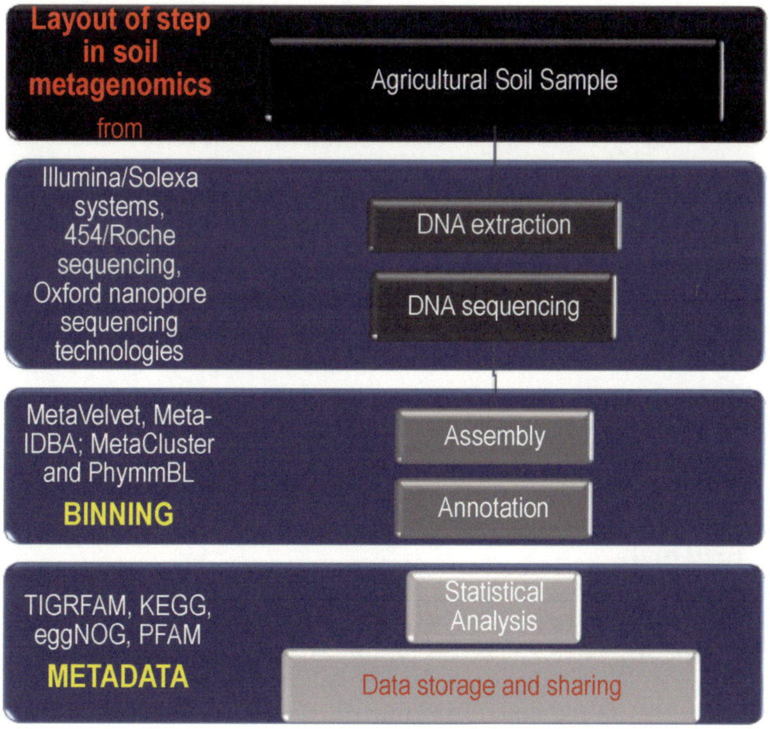

Figure 2.
The layout of metagenomics showing collection of samples from agricultural field and analysis.

6.1 Sample processing

Sampling is the first and crucial step. The extracted DNA must be of high quality for metagenomic library construction and sequencing. Further, fractionation or selective lysis is ideal for those communities which are linked to the plethora.

Fractionation should be examined for adequate target enrichment with little contamination.

6.2 Sequencing and assembly

Metagenomic sequencing significantly depends upon the sequencing platforms used. Nowadays, NGS techniques viz. Illumina/Solexa systems, 454/Roche sequencing, and Oxford nanopore sequencing technologies are being continuously used for metagenomic projects. Contigs are essential in getting the whole length sequence. So, assembly of short reads becomes key in metagenomics which may be accomplished by co-assembly and de novo assembly methods. On the flip side, the de novo assembly needs sophisticated computational tools and assemblers (e.g. MetaVelvet, and Meta-IDBA).

6.3 Binning and annotation

Binning shows the process of sorting of DNA in several groups of individual genomes.

In the very first step, binning explores the conserved nucleotide composition of genomes. Then, the DNA fragments are searched against a reference to bin the sequence. The binning algorithms use structure and similarity, like MetaCluster and PhymmBL. If the goal of the analysis of the reconstructed genome and large contigs, in this particular strategy, little length of contigs should be 30,000 bp or even longer. In future prediction of the assembled sequences, labelling is done while functional annotation includes mapping with an existing database. The sequences which cannot be mapped provide an endless amount of novelty in metagenomic samples. Several reference databases can be utilized to supply functional annotation viz. TIGRFAM, KEGG, eggNOG, PFAM, etc.

6.4 Statistical analysis and data sharing

Statistical assessment of the metagenomic data is vital for the exploration of the significance of the results. However, it must have appropriate experimental designs with proper replications. Metagenomic data sharing involves a great computational framework as well as a storage facility. Several of the centralized services have typical formats for recording and documenting experimental details.

7. Future road map

Robust extraction, as well as characterization of the DNA of soil microbiota through amplicon sequencing, has revolutionized the studies of ecology and environmental sciences. Essentially, the metagenomic evaluation of nucleic acids gives immediate access to the genomes of the uncultivated majority of underexploited microbial life. Accelerated by developments in sequencing technologies, microbiologists have found more novel species, genera, as well as genes from microorganisms. The unprecedented range of soil types continued exploration of a variety of agricultural and environmental features. The capacity to check out earth microbial communities with increasing capability has presumably the highest promise for answering numerous mysteries of the microbial world. Molecular methods, which include metagenomics, have revolutionized the analysis of microbial ecology.
We cannot link virtually all microorganisms to their metabolic roles within an earth community. Increased sequencing capability provided by high throughput

sequencing technologies has assisted characterize as well as quantify soil diversity. However, these methodologies are usually leveraged to process more samples at a reasonably shallow depth as compared to survey throughout the genomes from a single sample adequately. **Figure 3** describes the various application of metagenomics.

Along with higher diversity, methodological biases produce a considerable challenge for soil microbial characterization. These biases include soil sampling, DNA extraction, adsorption of nucleic acids to soil particles, contributions of extracellular DNA, sample planning, sequencing protocols, sequence analysis, and purposeful annotation. Since current sequencing technologies produce millions of reads, difficulties linked to interpreting these results can contribute to the problems encountered by microbial ecologists in determining the involvement of various microorganisms in the number of processes of soil. Without having a suitable benchmark methodology or dataset for verifying the fidelity of amplicon or perhaps metagenomic analyses, assessing whether the presence, as well as the activity of organisms, are adequately evaluated, is impossible. Furthermore, methodological limitations which might stop the detection of some active and abundant bacteria in soil could lead to the same essential amount of misinterpretation. No individual protocol would be seen as adequate in isolation of DNA. Likewise, the taxonomic and likely useful deciphering of the soil microbiota would critically gain from a blend of strategies.

Exact replicates are challenging to obtain due to soil microorganism compositional changes. An additional challenge would be that the total number of species that are in a single sample of soil is unfamiliar, with hugely varying estimates. One crucial very first step toward dealing with several of the problems experienced by soil microbiologists is actually to start developing a substantial catalog of all microbial community members and features for no less than one reference soil. Such a relatively comprehensive reference dataset would shed light on the as-yet-unknown design of a ground microbial species frequency distribution and might serve as an ultimate guide for assessing town composition switches across soil landscapes (i.e., beta diversity). Put simply, the scope of bias with any private strategy (i.e., a one-time DNA extraction method) might be explicitly driven by comparing extraction

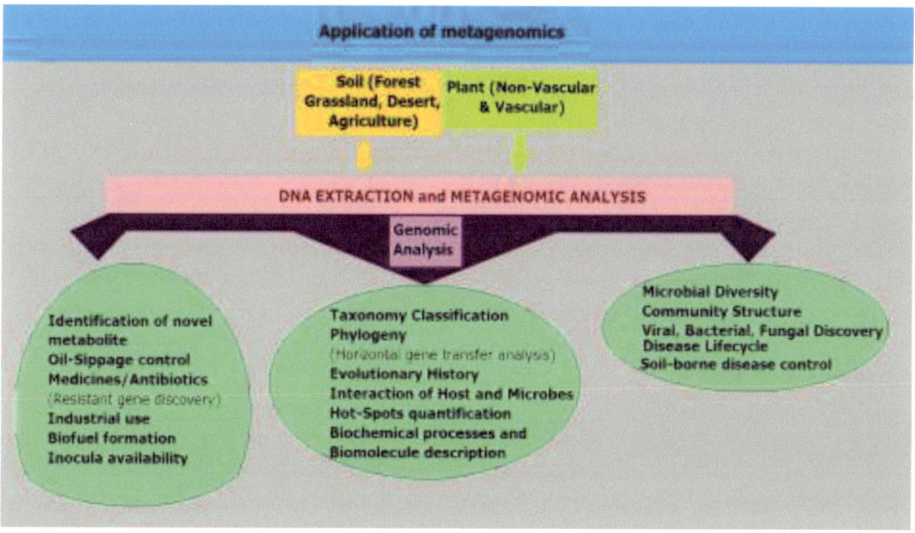

Figure 3.
A brief account of applications of metagenomics in different fields.

strategies coupled with detailed characterization of the selected reference soil. For instance, the isolation, as well as characterization of cells via single-cell genomics, can assist target phylogenetically analysis. Coupled with extensive DNA based characterization of the collected guide soil microbial diversity, this specific research initiative should ideally assess several levels of gene expression, at the amount of RNA (metatranscriptomics), proteins (metaproteomics), and also metabolites (metametabolomics). By identifying the way a reference soil is structured, both temporally and spatially, the info from this coordinated effort might help supply missing links between typical soil analyses as well as the underlying composition of soil microbial communities.

An in-depth exploration of single guide soil should involve experiments much beyond the typical metagenomic analyses applied to soil samples. Instead, this effort is going to require considerable benchmarking of the sampling technique itself, which is connected to identifying a suitable resource website. Such an endeavor would call for a coordinated inter-disciplinary consortium of knowledge spanning chemistry, soil physics, biochemistry, microbiology, and bioinformatics. The outcomes of the effort can develop an objective foundation for creating standardized protocols for ongoing and future soil microbiological investigations.

Author details

Prashant Kaushik[1,2*], Opinder Singh Sandhu[3], Navjot Singh Brar[4], Vivek Kumar[5], Gurdeep Singh Malhi[5], Hari Kesh[6] and Ishan Saini[7]

1 Instituto de Conservación y Mejora de la Agrodiversidad Valenciana, Universitat Politècnica de València, Valencia, Spain

2 Nagano University, Nagano, Japan

3 Department of Soil Science, Punjab Agricultural University, Ludhiana, India

4 Department of Vegetable Science, Punjab Agricultural University, Ludhiana, India

5 Department of Agronomy, Punjab Agricultural University, Ludhiana, India

6 Department of Plant Breeding, CCS Haryana Agricultural University, Hisar, India

7 Department of Botany, Kurukshetra University, Kurukshetra, Haryana, India

*Address all correspondence to: prakau@doctor.upv.es

References

[1] Berendsen RL, Pieterse CM, Bakker PA. The rhizosphere microbiome and plant health. Trends in Plant Science. 2012;**17**:478-486

[2] Garbeva P, Postma J, van Veen JA, van Elsas JD. Effect of above-ground plant species on soil microbial community structure and its impact on suppression of *Rhizoctonia solani* AG3. Environmental Microbiology. 2006;**8**:233-246

[3] Zhao M, Cong J, Cheng J, Qi Q, Sheng Y, Ning D, et al. Soil microbial community assembly and interactions are constrained by nitrogen and phosphorus in broadleaf forests of southern China. Forests. 2020;**11**(3):285

[4] Schloss PD, Handelsman J. Toward a census of bacteria in soil. PLOS Computational Biology. 2006;**2**:e92

[5] Uroz S, Buée M, Deveau A, Mieszkin S, Martin F. Ecology of the forest microbiome: Highlights of temperate and boreal ecosystems. Soil Biology and Biochemistry. 2016;**103**:471-488

[6] Fierer N, Leff JW, Adams BJ, Nielsen UN, Bates ST, Lauber CL, et al. Cross-biome metagenomic analyses of soil microbial communities and their functional attributes. Proceedings of the National Academy of Sciences of the United States of America. 2012;**109**:21390-21395

[7] Saini I, Aggarwal A. Kaushik P, Influence of biostimulants on important traits of *Zinnia elegans* Jacq. Under open field conditions. International Journal of Agronomy. 2019;**2019**:3082967

[8] Saini I, Aggarwal A, Kaushik P. Inoculation with mycorrhizal fungi and other microbes to improve the morpho-physiological and floral traits of *Gazania rigens* (L.) Gaertn. Agriculture. 2019;**9**:51

[9] Jia Y, Whalen JK. A new perspective on functional redundancy and phylogenetic niche conservatism in soil microbial communities. Pedosphere. 2020;**30**(1):18-24

[10] Zhang L, Jing Y, Xiang Y, Zhang R, Lu H. Responses of soil microbial community structure changes and activities to biochar addition: A meta-analysis. Science of the Total Environment. 2018;**643**:926-935

[11] Wu Z, Haack SE, Lin W, Li B, Wu L, Fang C, et al. Soil microbial community structure and metabolic activity of *Pinus elliottii* plantations across different stand ages in a subtropical area. PLOS One. 2015;**10**(8):e0135354

[12] Mitchell PJ, xSimpson AJ, Soong R, Simpson MJ. Shifts in microbial community and water-extractable organic matter composition with biochar amendment in a temperate forest soil. Soil Biology and Biochemistry. 2015;**81**:244-254

[13] An SS, Cheng Y, Huang YM, Liu D. Effects of revegetation on soil microbial biomass, enzyme activities, and nutrient cycling on the Loess Plateau in China. Restoration Ecology. 2012;**21**:600-607

[14] Jia G, Cao J, Wang C, Wang G. Microbial biomass and nutrients in soil at the different stages of secondary forest succession in Ziwulin, Northwest China. Forest Ecology and Management. 2005;**217**:117-125

[15] Sessitsch A, Weilharter A, Gerzabek MH, Kirchmann H, Kandeler E. Microbial population structures in soil particle size fractions of a long-term fertilizer field experiment. Applied and Environmental Microbiology. 2001;**67**:4215-4224

[16] Suyal DC, Joshi D, Debbarma P, Soni R, Das B, Goel R. Soil metagenomics: Unculturable microbial diversity and its function. In: Varma A, Choudhary D, editors. Mycorrhizosphere and Pedogenesis. Springer: Singapore; 2019. pp. 355-362

[17] Nair GR, Raja SSS. Decoding complex soil microbial communities through new age "Omics". Journal of Microbial and Biochemical Technology. 2017;**9**(6):301-309

[18] Blagodatskaya E, Kuzyakov Y. Active microorganisms in soil: Critical review of estimation criteria and approaches. Soil Biology and Biochemistry. 2013;**67**:192-211

[19] Qaisrani MM, Zaheer A, Mirza MS, Naqqash T, Qaisrani TB, Hanif MK, et al. A comparative study of bacterial diversity based on culturable and culture-independent techniques in the rhizosphere of maize (*Zea mays* L.). Saudi Journal of Biological Sciences. 2019;**26**(7):1344-1351

[20] Long Y, Jiang J, Hu X, Zhou J, Hu J, Zhou S. Actinobacterial community in Shuanghe Cave using culture-dependent and -independent approaches. World Journal of Microbiology and Biotechnology. 2019;**35**:153

[21] Xiong W, Zhao Q, Zhao J, Xun W, Li R, Zhang R, et al. Different continuous cropping spans significantly affect microbial community membership and structure in a vanilla-grown soil as revealed by deep pyrosequencing. Microbiology Ecology. 2014;**70**:209-218

[22] Wang C, Zheng MM, Hu AY, Zhu CQ, Shen RF. Diazotroph abundance and community composition in an acidic soil in response to aluminum-tolerant and aluminum-sensitive maize (*Zea mays* L.) cultivars under two nitrogen fertilizer forms. Plant and Soil. 2018;**424**:463-478

[23] Nesme J, Achouak W, Agathos SN, Bailey M, Baldrian P, Brunel D, et al. Back to the future of soil metagenomics. Frontiers in Microbiology. 2016;**7**:73

[24] Berg G. Plant–microbe interactions promoting plant growth and health: Perspectives for controlled use of microorganisms in agriculture. Applied Microbiology and Biotechnology. 2009;**84**:11-18

[25] Ventura M, Turroni F, Canchaya C, Vaughan EE, O'Toole PW, van Sinderen D. Microbial diversity in the human intestine and novel insights from metagenomics. Frontiers in Bioscience. 2009;**14**(1):3214-3863

[26] Smith RJ, Jeffries TC, Roudnew B, Fitch AJ, Seymour JR, Delpin MW, et al. Metagenomic comparison of microbial communities inhabiting confined and unconfined aquifer ecosystems. Environmental Microbiology. 2011;**14**(1):240-253

[27] Bertagnolli AD, Stewart FJ. Microbial niches in marine oxygen minimum zones. Nature Reviews Microbiology. 2018;**16**:723-729

[28] Hemme CL, Deng Y, Gentry TJ, Fields MW, Wu L, Barua S, et al. Metagenomic insights into evolution of a heavy metal-contaminated groundwater microbial community. The ISME Journal. 2010;**4**:660-672

[29] Silver S, Phung LT. Bacterial heavy metal resistance: New surprises. Annual Review of Microbiology. 1996;**50**:753-789

[30] Sabale SN, Suryawanshi PP, Krishnaraj PU. Soil metagenomics: Concepts and applications. In: Metagenomics-Basics, Methods and Applications. London, United Kingdom: IntechOpen; 2019

[31] Saini I, Yadav K, Esha E, Aggarwal A. Effect of bioinoculants

on morphological and biochemical parameters of *Zinnia elegans* Jacq. Journal of Applied Horticulture. 2017;**19**(2):167-172

[32] Saini I, Yadav K, Aggarwal A. Response of arbuscular mycorrhizal fungi along with *Trichoderma viride* and *Pseudomonas fluorescens* on the growth, biochemical attributes and vase life of *Chrysanthemum indicum*. Journal of Environmental Biology. 2019c;**40**:183-191

[33] Abawi GS, Widmer TL. Impact of soil health management practices on soilborne pathogens, nematodes and root diseases of vegetable crops. Applied Soil Ecology. 2000;**15**(1):37-47

[34] Dumbrell A, Nelson M, Helgason T, Dytham C, Fitter AH. Relative roles of niche and neutral processes in structuring a soil microbial community. The ISME Journal. 2010;**4**:337-345

[35] Lennon JT, Aanderud ZT, Lehmkuhl BK, Schoolmaster DR Jr. Mapping the niche space of soil microorganisms using taxonomy and traits. Ecology. 2012;**93**(8):1867-1879

[36] Green JL, Bohannan BJM, Whitaker RJ. Microbial biogeography: From taxonomy to traits. Science. 2008;**320**:1039-1043

[37] Wallenstein MD, Hall EK. A trait-based framework for predicting when and where microbial adaptation to climate change will affect ecosystem functioning. Biogeochemistry. 2011;**109**:35-47

[38] Heisler C, Kaiser E. Influence of agricultural traffic and crop management on collembola and microbial biomass in arable soil. Biology and Fertility of Soils. 1995;**19**:159-165

[39] Ponge JF, Pérès G, Guernion M, Ruiz-Camacho N, Cortet J, Pernin C, et al. The impact of agricultural practices on soil biota: A regional study. Soil Biology and Biochemistry. 2013;**67**:271-284

[40] Yan S, Singh AN, Fu S, Liao C, Wang S, Li Y, et al. A soil fauna index for assessing soil quality. Soil Biology and Biochemistry. 2012;**47**:158-165

[41] Byers HG, Kellogg CE, Anderson MS, Thorp J. Formation of soil. In: Soils and Men: U.S. Department Agriculture Yearbook. Washington, D.C., USA. 1938. pp. 948-992

[42] Humphreys CP, Franks PJ, Rees M, Bidartondo MI, Leake JR, Beerling DJ. Mutualistic mycorrhiza-like symbiosis in the most ancient group of land plants. Nature Communications. 2010;**1**:103

[43] Louca S, Polz MF, Mazel F, Albright MBN, Huber JA, O'Connor MI, et al. Function and functional redundancy in microbial systems. Nature Ecology and Evolution. 2018;**2**:936-943

[44] Marco DE, Abram F. Editorial: Using genomics, metagenomics and other "omics" to assess valuable microbial ecosystem services and novel biotechnological applications. Frontiers in Microbiology. 2019;**10**:151

[45] Hugenholtz P, Tyson G. Metagenomics. Nature. 2008;**455**: 481-483

[46] Tyson G, Chapman J, Hugenholtz P, Allen EE, Ram RJ, Richardson PM, et al. Community structure and metabolism through reconstruction of microbial genomes from the environment. Nature. 2004;**428**:37-43

[47] National Research Council of the National Academies. The dawning of a new microbial age. In: The New Science of Metagenomics: Revealing the Secrets of our Microbial Planet. Washington, DC: The National Academies Press; 2007. p. 2

[48] Gill SR, Pop M, DeBoy RT, Eckburg PB, Eckburg PB, Turnbaugh PJ, et al. Metagenomic analysis of the human distal gut microbiome. Science. 2006;**312**:1355-1359

[49] Rosano GL, Ceccarelli EA. Recombinant protein expression in *Escherichia coli*: Advances and challenges. Frontiers in Microbiology. 2014;**5**:172

[50] Elena C, Ravasi P, Castelli M, Peiru S, Menzella H. Expression of codon optimized genes in microbial systems: Current industrial applications and perspectives. Frontiers in Microbiology. 2014;**5**:21

[51] Rashid M, Stingl U. Contemporary molecular tools in microbial ecology and their application to advancing biotechnology. Biotechnology Advances. 2015;**33**(8):1755-1773

[52] Zielińska S, Kidawa D, Stempniewicz L, Łoś M, Łoś JM. Environmental DNA as a valuable and unique source of information about ecological networks in Arctic terrestrial ecosystems. Environmental Reviews. 2017;**25**(3):282-291

[53] McGee KM, Robinson CV, Hajibabaei M. Gaps in DNA-based biomonitoring across the globe. Frontiers in Ecology and Evolution. 2019;**7**:1-7

[54] Jovel J, Patterson J, Wang W, Hotte N, O'Keefe S, Mitchel T, et al. Characterization of the gut microbiome using 16S or shotgun metagenomics. Frontiers in Microbiology. 2016;**7**:459

[55] Sboner A, Mu XJ, Greenbaum D, Auerbach RK, Gerstein MB. The real cost of sequencing: Higher than you think! Genome Biology. 2011;**12**:125

[56] Knights D, Kuczynski J, Charlson E, Zaneveld F, Mozer MC, Collman RG, et al. Bayesian community-wide culture-independent microbial source tracking. Nature Methods. 2011;**8**:761-763

[57] Shamim K. Metagenomic approach to screen protease encoding genes from Coastal Environment of Goa [PhD thesis]. Goa, India: Goa University; 2017

[58] Singh A, Kumar M, Verma S, Choudhary P, Chakdar H. Plant microbiome: Trends and prospects for sustainable agriculture. In: Plant Microbe Symbiosis. New York City, USA: Springer; 2020. pp. 129-151

[59] Morgan HH. Investigating the effect of leaf removal on the grape-associated microbiome through culture-dependent and–independent approaches [PhD thesis]. Stellenbosch: Stellenbosch University; 2016

[60] Mocali S, Benedetti A. Exploring research frontiers in microbiology: The challenge of metagenomics in soil microbiology. Research in Microbiology. 2010;**161**:497-505

[61] Zarraonaindia I, Smith DP, Gilbert JA. Beyond the genome: Community-level analysis of the microbial world. Biology and Philosophy. 2013;**28**:261-282

[62] Fierer N, Jackson RB. The diversity and biogeography of soil bacterial communities. Proceedings of the National Academy of Sciences of the United States of America. 2006;**103**(3):626-631

[63] Gonzalez A, King A, Robeson MS II, Song S, Shade A, Metcalf JL, et al. Characterizing microbial communities through space and time. Current Opinion in Biotechnology. 2012;**23**(3):431-436

[64] DeLong EF, Pace NR. Environmental diversity of bacteria and archaea. Systematic Biology. 2001;**50**:470-478

[65] Liao K, Bai Y, Huo Y, Jian Z, Hu W, Zhao C, et al. Integrating microbial biomass, composition and function to discern the level of anthropogenic activity in a river ecosystem. Environment International. 2018;**116**:147-155

[66] Andersen R, Chapman SJ, Artz RRE. Microbial communities in natural and disturbed peatlands: A review. Soil Biology and Biochemistry. 2013;**57**:979-994

[67] Szynkowska MI, Pawlaczyk A, Lesniewska E, Paryjczak T. Toxic metal distribution in rural and urban soil samples affected by industry and traffic. Polish Journal of Environmental Studies. 2009;**18**:1141-1150

[68] Wuana RA, Okieimen FE. Heavy metals in contaminated soils: A review of sources, chemistry, risks and best available strategies for remediation. International Scholarly Research Notices. 2011;**20**:402647

[69] Keller M, Zengler K. Tapping into microbial diversity. Nature Reviews Microbiology. 2004;**2**:141-150

[70] Prosser JI, Bohannan BJ, Curtis TP, Ellis RJ, Firestone MK, Freckleton RP, et al. The role of ecological theory in microbial ecology. Nature Reviews Microbiology. 2007;**5**:384-392

[71] Chang H-X, Haudenshield JS, Bowen CR, Hartman GL. Metagenome-wide association study and machine learning prediction of bulk soil microbiome and crop productivity. Frontiers in Microbiology. 2017;**8**:519

[72] Kielak A, Pijl AS, Van Veen JA, Kowalchuk GA. Phylogenetic diversity of acidobacteria in a former agricultural soil. The ISME Journal. 2009;**3**:378-382

[73] Carbonetto B, Rascovan N, Álvarez R, Mentaberry A, Vázquez MP. Structure, composition and metagenomic profile of soil microbiomes associated to agricultural land use and tillage systems in Argentine Pampas. PLOS One. 2014;**9**(6):e99949

[74] Souza RC, Cantão ME, Vasconcelos ATR, Nogueira MA, Hungria M. Soil metagenomics reveals differences under conventional and no-tillage with crop rotation or succession. Applied Soil Ecology. 2013;**72**:49-61

[75] Jongman M, Carmichael PC, Bill M. Technological advances in phytopathogen detection and metagenome profiling techniques. Current Microbiology. 2020;**77**:675-681

[76] Chalupowicz L, Dombrovsky A, Gaba V, Luria N, Reuven M, Beerman A, et al. Diagnosis of plant diseases using the Nanopore sequencing platform. Plant Pathology. 2019;**68**(2):229-238

[77] Yang Y, Zhang S, Li N, Chen H, Jia H, Song X, et al. Metagenomic insights into effects of wheat straw compost fertiliser application on microbial community composition and function in tobacco rhizosphere soil. Science Reports. 2019;**9**:6168

[78] Karaca M, Ince AG. Conservation of biodiversity and genetic resources for sustainable agriculture. In: Farooq M, Pisante M, editors. Innovations in Sustainable Agriculture. Cham: Springer; 2019. pp. 363-410

[79] Teuber S, Ahlrichs J, Henkner J, Knopf T, Kühn P, Scholten T. Soil cultures—The adaptive cycle of agrarian soil use in Central Europe: An interdisciplinary study using soil scientific and archaeological research. Ecology and Society. 2017;**22**(4):13

[80] Zamioudis C, Pieterse CMJ. Modulation of host immunity by beneficial microbes. Molecular Plant-Microbe Interactions. 2012;**25**:139-150

[81] De Coninck B, Timmermans P, Vos C, Cammue BPA, Kazan K. What lies beneath: Belowground defense strategies in plants. Trends in Plant Science. 2015;**20**(2):91-101

[82] Jansson JK, Hofmockel KS. The soil microbiome—From metagenomics to metaphenomics. Current Opinion in Microbiology. 2018;**43**:162-168

[83] Arora NK, Fatima T, Mishra I, Verma M, Mishra J, Mishra V. Environmental sustainability: Challenges and viable solutions. Environmental Sustainability. 2018;**1**:309-340

[84] Cruz AF, Ishii T. Arbuscular mycorrhizal fungal spores host bacteria that affect nutrient biodynamics and biocontrol of soil-borne plant pathogens. Biology Open. 2011;**1**:52-57

[85] Takahashi JS, Pinto LP, Vitaterna MH. Forward and reverse genetic approaches to behavior in the mouse. Science. 1994;**264**:1724-1733

[86] Schlaeppi K, Bulgarelli D. The plant microbiome at work. Molecular Plant-Microbe Interactions. 2015;**28**(3):212-217

[87] Bulgarelli D, Schlaeppi K, Spaepen S, van Themaat EVL, Schulze-Lefert P. Structure and functions of the bacterial microbiota of plants. Annual Review of Plant Biology. 2013;**64**:807-838

[88] Kaushik P. Classification of Indian States and Union Territories based on their Soil Macronutrient and Organic Carbon Profiles. BioRxiv. 2020. DOI: 10.1101/2020.02.10.930586

[89] Poretsky R, Rodriguez-R LM, Luo C, Tsementzi D, Konstantinidis KT. Strengths and limitations of 16S rRNA gene amplicon sequencing in revealing temporal microbial community dynamics. PLOS One. 2014;**9**:e93827

[90] Tamaki H, Wright CL, Li X, Lin Q, Hwang C, Wang S, et al. Analysis of 16S rRNA amplicon sequencing options on the Roche/454 next-generation titanium sequencing platform. PLOS One. 2011;**6**(9):e25263

[91] Ranjan R, Rani A, Metwally A, McGee HS, Perkins DL. Analysis of the microbiome: Advantages of whole genome shotgun versus 16S amplicon sequencing. Biochemical and Biophysical Research Communications. 2016;**469**:967-977

[92] Delmont TO, Robe P, Clark I, Simonet P, Vogel TM. Metagenomic comparison of direct and indirect soil DNA extraction approaches. Journal of Microbiological Methods. 2011;**86**(3):397-400

[93] Chistoserdova L. Recent progress and new challenges in metagenomics for biotechnology. Biotechnology Letters. 2010;**32**:1351-1359

[94] Sjöling S, Cowan DA. Metagenomics: Microbial community genomes revealed. In: Margesin R, Schinner F, Marx J-C, Gerday C, editors. Psychrophiles: From Biodiversity to Biotechnology. Germany: Springer-Verlag, Berlin; 2008. pp. 313-332

[95] Sugiyama A, Vivanco JM, Jayanty SS, Manter DK. Pyrosequencing assessment of soil microbial communities in organic and conventional potato farms. Plant Disease. 2010;**94**(11):1329-1335

[96] Lim YW, Kim BK, Kim C, Jung HS, Kim BS, Lee JH, et al. Assessment of soil fungal communities using pyrosequencing. Journal of Microbiology. 2010;**48**:284-289

[97] Nacke H, Thürmer A, Wollherr A, Will C, Hodac L, Herold N, et al. Pyrosequencing-based assessment of

bacterial community structure along different management types in German forest and grassland soils. PLOS One. 2011;**6**(2):e17000

[98] Xu L, Ravnskov S, Larsen J, Nilsson RH, Nicolaisen M. Soil fungal community structure along a soil health gradient in pea fields examined using deep amplicon sequencing. Soil Biology and Biochemistry. 2012;**46**:26-32

[99] Brown SP, Callaham MA Jr, Oliver AK, Jumpponen A. Deep Ion Torrent sequencing identifies soil fungal community shifts after frequent prescribed fires in a southeastern US forest ecosystem. FEMS Microbiology Ecology. 2013;**86**(3):557-566

[100] Rodrigues VD, Torres TT, Ottoboni LMM. Bacterial diversity assessment in soil of an active Brazilian copper mine using high-throughput sequencing of 16S rDNA amplicons. Antonie Van Leeuwenhoek. 2014;**106**(5):879-890

[101] Ligi T, Oopkaup K, Truu M, Preem JK, Nõlvak H, Mitsch WJ, et al. Characterization of bacterial communities in soil and sediment of a created riverine wetland complex using high-throughput 16S rRNA amplicon sequencing. Ecological Engineering. 2014;**72**:56-66

[102] Mwaikono KS, Maina S, Sebastian A, Kapur V, Gwakisa P. 16S rRNA amplicons survey revealed unprecedented bacterial community in solid biomedical wastes. American Journal of Microbiological Research. 2015;**3**(4):135-143

[103] Finley SJ, Pechal JL, Benbow ME, Robertson BK, Javan GT. Microbial signatures of cadaver gravesoil during decomposition. Microbial Ecology. 2016;**71**(3):524-529

[104] Jenkins JR, Viger M, Arnold EC, Harris ZM, Ventura M, Miglietta F, et al. Biochar alters the soil microbiome and soil function: Results of next-generation amplicon sequencing across Europe. GCB Bioenergy. 2017;**9**:591-612

[105] Ros M, Blaya J, Baldrian P, Bastida F, Richnow HH, Jehmlich N, et al. In vitro elucidation of suppression effects of composts to soil-borne pathogen *Phytophthora nicotianae* on pepper plants using 16S amplicon sequencing and metaproteomics. Renewable Agriculture and Food Systems. 2018;**35**(2):1-9

[106] De Mandal S, Mathipi V, Muthukumaran RB, Gurusubramanian G, Lalnunmawii E, Kumar NS. Amplicon sequencing and imputed metagenomic analysis of waste soil and sediment microbiome reveals unique bacterial communities and their functional attributes. Environmental Monitoring and Assessment. 2019;**191, 12**:778

[107] Manzotti A, Bergna A, Burow M, Jørgensen HJL, Cernava T, Berg G, et al. Insights into the community structure and lifestyle of the fungal root endophytes of tomato by combining amplicon sequencing and isolation approaches with phytohormone profiling. FEMS Microbiology Ecology. 2020;**96**(5):fiaa052

[108] Daniel R. The metagenomics of soil. Nature Reviews Microbiology. 2005;**3**:470-478

[109] Ferrer M, Beloqui A, Timmis KN, Golyshin PN. Metagenomics for mining new genetic resources of microbial communities. Journal of Molecular Microbiology and Biotechnology. 2009;**16**:109-123

[110] Gabor EM, Alkema WB, Janssen DB. Quantifying the

accessibility of the metagenome by random expression cloning techniques. Environmental Microbiology. 2004;**6**:879-886

[111] Craig JW, Chang F-Y, Kim JH, Obiajulu SC, Brady SF. Expanding small-molecule functional metagenomics through parallel screening of broad-host-range cosmid environmental DNA libraries in diverse Proteobacteria. Applied and Environmental Microbiology. 2010;**76**:1633-1641

Chapter 2

Arbuscular Mycorrhizal (AM) Fungi as a Tool for Sustainable Agricultural System

Kavita Chahal, Vaishali Gupta, Naveen Kumar Verma, Anand Chaurasia and Babita Rana

Abstract

A sustainable agriculture is a type of agriculture that focuses on producing long-term crops and livestock without having any adverse effect on the environment. However, agricultural malpractices like excessive use of chemical fertilizers and pesticides, as well as climate change have aggravated the effects of biotic and abiotic stresses on crop productivity. These led to the degradation of ecosystem, leaving bad impacts on the soil qualities and water body environment. As an alternative to the rising agricultural energy, the use of Vesicular– Arbuscular Mycorrhizae (AM) may be a better option. Being natural root symbionts, AM provide essential inorganic nutrients to host plants, thereby improving its growth and yield even under stressed conditions. AM fungi can also potentially strengthen the adaptability of a plant to the changing environment, as a bio-fertilizer. The chapter provides a comprehensive up-to-date knowledge on AM fungi as a tool for sustainable agricultural system. Thus, further research focusing on the AM -mediated promotion of crop quality and productivity is needed.

Keywords: vesicular–arbuscular mycorrhizae, sustainable, symbionts, productivity

1. Introduction

A potential solution to enable agricultural systems to feed a growing population within the changing environmental conditions is a sustainable agriculture, that is based on an understanding of the society's present food and textile needs, as well as on the ecosystem services. A special attention must be needed towards the study of the ability of symbiotic relationship among the actinorhizal plants and microbes, so as to overcome the problems of deforestation and the increasing cost of nitrogenous fertilizers [1].

For increasing the sustainability of agriculture, among various other methods, the better option is the use of natural root symbiont, Arbuscular Mycorrhizae (AM). As compared to conventional agriculture, the soil conditions are likely to be more favorable to AM fungi in a sustainable agriculture [2–4].

The AM fungi have been found to be associated with more than 80% of land plants, liverworts, ferns, gymnosperms, angiosperms and grasses, and are widely distributed in natural and agricultural environments. Hence, for crop and biomass production, these symbiotic associations are very important, and they are receiving

IntechOpen

considerable attention in forestry and agriculture. Therefore, AM fungi are commonly known as bio-fertilizers. Moreover, these natural root symbionts help their host plants to grow vigorously under stressful conditions like drought, salinity, metals, and extreme temperatures. The mechanism behind is a series of complex communication events between the plant and the fungus that lead to increased photosynthetic rate and increased water uptake [5–7]. The AM fungi also assists in the regulation of metabolic pathways in plants.

2. Sustainable agricultural system and its benefits

The best agricultural system is the one that find a good balance between the need for food production and the preservation of the ecological system within the environment. Also, the agriculture that focuses on producing long-term crops and livestock with minimum adverse effects on the environment is called as sustainable agriculture [8]. Besides the production of food, some of the various objectives of sustainable agriculture include reducing the use of fertilizers and pesticides, promoting biodiversity in crops grown and the ecosystem, and conserving water. It also aimed at maintaining the economic stability of farms and improving farming techniques and quality of a farmer's life [9].

The various benefits of sustainable agriculture can be divided into human health benefits and environmental benefits. Regarding human health, crops grown through sustainable agriculture are better for people. People are not becoming ill by consuming synthetic materials present in chemical pesticides and fertilizers. In addition, these crops are also more nutritious because of their more natural production. Its positive impacts of the environment include use of less percentage of energy per unit of crop yield as compared to industrialized agriculture [10, 11]. This minimizes the release of harmful chemicals and thereby reduces pollution of environment. Other benefits to the environment are maintenance of soil quality, reduction in soil degradation and erosion, and saving water. It also increases biodiversity of the area by providing a variety of organisms surrounded with a healthy and natural environment (**Figure 1**) [12–14].

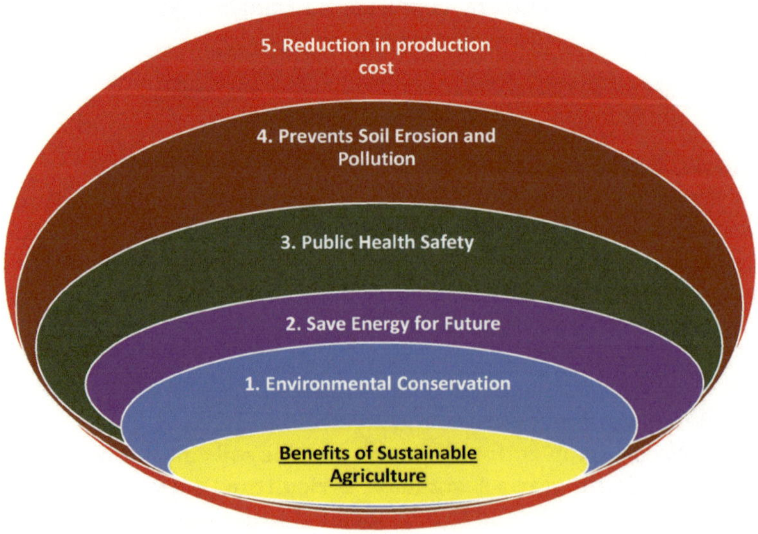

Figure 1.
Benefits of sustainable agriculture

3. Mycorrhizal association

The symbiotic association between a fungus and a root of higher plant is called as Mycorrhiza [15]. From this association, both of the partners, the host plant and the fungal member are benefited potentially [16]. There are several benefits provided by the Mycorrhizal fungi to the host plant species. Some of them are increased nutrient uptake, production of growth promoting substances, tolerance to drought, salinity and synergistic interactions with other beneficial microorganisms [17, 18].

4. Arbuscular mycorrhizae

Endomycorrhiza i.e. Arbuscular Mycorrhizal (AM) symbiosis is formed by approximately 80% of all terrestrial plant species. Even the roots of some aquatic plants are colonized by AM fungi [19, 20]. AM fungi belong to the class Zygomycetes, order Endogonales, family Endogonaceae, and phylum Glomeromycota. The mycorrhizal associations are formed by the six genera of fungi belonging to Endogonaceae. These are Glomus, Gigaspora, Acaulospora, Entrophospora Sclerocystis and Scutellospora. Their common characteristic are spores and sporocarps which are formed mostly in the soil surrounding the roots and rarely inside the roots [21].

The most visible AM structure is the hyphal network. Hyphae are thin from 2 μm in diameter to >20 μm, hollow tubes of fungi having only few cross walls and distinct angular projections [22]. In search of the roots of host plants, these tubes originally grow from fungal spores, extending short distances into the soil.

Hyphae that penetrates a host root form a structure called an appressorium. It penetrates the cell wall of the root by mechanical pressure or through the enzymes that degrade the cell wall. Hyphae that enter host roots through these infection points can form networks both inside the root and throughout the soil surrounding the root. As the name suggests, the AM fungus colonizes the root cortex forming a mycelial network and characteristic bladder-like structures called as vesicles and branched finger-like hyphae called as arbuscules. Arbuscules are short-lived structures meant for nutrient transfer and absorptive function. The hyphal branch that penetrates the plant cell wall forms the arbuscules trunk. This arbuscule trunk branches repeatedly and is surrounded by the plasma membrane. The terminal swellings of the hyphae forms vesicles on both intercellular and intracellular surfaces, and have storage as function [23, 24].

4.1 As a tool for sustainable agriculture

4.1.1 Benefits from tripartite relationship

In the mutualistic association, the plant provides the fungus with photosynthetically derived carbohydrate, while the fungus supplies the plant roots with nutrients. Also, in this symbiotic association, there is a third component i.e., a bacterium that seems to be having a loose or tight association with the plants and the mycorrhizal fungi and play an important role in mycorrhizal function. So, there is a tripartite relationship among host plant, AM fungi and bacteria. This bacterium has been termed as 'helper bacteria' because it supports mycorrhizal establishment [25].

For the establishment of a symbiotic relationship with the nitrogen fixing rhizobium bacteria, the AM fungi releases a 'myc factor' which is a diffusional factor responsible for activating the nodulation factor's inducible gene MtEnod11. This gene is involved in establishing symbiotic relationship with the nitrogen fixing rhizobial bacteria [26–28]. Under natural conditions, this bacterium live in the cytoplasm as endobacteria or colonize the surface of extraradical hyphae [29].

4.1.2 Natural growth regulators

AM fungi are used as bio-inoculants, and as prominent natural growth regulators in sustainable crop productivity. Also, the stomatal conductance, leaf water potential, relative water content, photosystem II efficiency, and carbon dioxide assimilation are improved by AM inoculation that contribute greatly to organic culturing for growth promotion and yield maximization [30–32].

4.1.3 Bio-fertilizer

For fulfilling the fertilizer requirements of plants in areas of marginal fertility and to reduce the harmful effects of chemical fertilizer, AM have a potential use as a biofertilizer. Bio- fertilizers are a mixture of naturally occurring substances for improving soil fertility [33]. Various problems and damaging impact on the quality of food products, soil health, and air and water systems are associated with the continuous use of inorganic fertilizers, herbicides, and fungicides. Reports showed the AM can possibly lower down the use of chemical fertilizers up to 50% for best agricultural production [34].

4.1.4 Plant yield

AM Fungi can also have potential to enhance the dietary quality of crops and to increase the levels of secondary metabolites and production of carotenoids and certain volatile compounds. There are reports that showed beneficial effects of AM fungi *Glomus versiforme* on the increased contents of sugars, organic acids, vitamin C, flavonoids, and minerals resulting in enhanced citrus fruit quality. It enhances plant yield for a healthy food production chain by increasing the accumulation of anthocyanins, chlorophyll, carotenoids, total soluble phenolics, tocopherols, and various mineral nutrients. The field production of maize, yam, and potato, has been significantly increased using AM fungi [34–36].

4.1.5 Mineral nutrition cycle

The performance of most agricultural crops becomes better and is more productive in the presence of AM fungi. Mycorrhiza develops symbiosis with roots to obtain essential nutrients from the host plant and consequently provide mineral nutrients in return, for example, N, P, K, Ca, Zn, and S. This symbiosis increases the micronutrient uptake and growth of their plant host [37]. It has an important function in promoting the mineral cycling by maintaining an efficient and closed nutrient cycle of natural ecosystems, thereby changing the ecology of surrounding environment. An increase in the accumulation of biomass is also observed by the inoculation of AM fungi. This is because AM fungi increases the concentration of various macro-nutrients significantly, leading to increased photosynthate production [38, 39]. Thus, even under inappropriate conditions it provides nutritional support to the plants.

4.1.6 Transport of phosphorus and nitrogen to plants

The AM fungi are important to their hosts as they enhance the ability of plants to absorb phosphorus from soil, which is relatively inaccessible to the plants. The arbuscules of the fungi assist in exchange of inorganic minerals and the compounds of carbon and phosphorus imparting a considerable strength to host plants. Therefore, it significantly boosts the phosphorus concentration in both root and shoot systems. Also, under phosphorus-limited conditions, the association improves phosphorus supply to the infected roots of host plants. For phosphorus uptake, the crops that are poor at seeking out nutrients in the soil are dependent on AM fungi. It has significant effects on different plant communities, particularly on invasive plants and the fungal-mediated transport of phosphorus and nitrogen to plants [40–42].

4.1.7 Phyto-availability of micronutrients

A part from the macronutrients, AM fungi association has been reported to increase the phyto-availability of micronutrients like zinc and copper. Also, it helps the plants to take up nutrients from the nutrient-deficient soils. It is also responsible for the uptake of almost all essential nutrients, specifically phosphate, in plants. It was also reported to increase the absorption of trace elements, such as boron and molybdenum [43–45].

On the other hand, it also decreases the uptake of sodium and chlorine thereby stimulating the plant growth. Increased nitrogen content in plants evidently results in higher chlorophyll contents that can effectively trap nitrogen. Maintenance of calcium ion and sodium ion ratio helps improve the overall plant performance. It also improves the surface absorbing capability of host roots [46–48].

Some examples of enhancement of mineral nutrition:

- In mycorrhizal chickpea, improved growth and levels of protein, iron, and zinc were found [49].

- In the mycorrhizal roots of *Lotus japonicus*, an enhanced activity of a potassium transporter was reported [50].

- AM fungi when inoculated in tomato plants have shown increased leaf area, and nitrogen, potassium, calcium, and phosphorus contents, showing an increased plant growth [51, 52].

- In *Pelargonium graveolens* L., mycorrhizal symbiosis increased the concentrations of Nitrogen, Phosphorus, and Iron under drought stress [53].

- In *Euonymus japonica,* improved levels of P, Ca, and K under salinity stress due to instant fungus attachment were reported. In another study, AMF-inoculated Pistachio plants exhibited high levels of P, K, Zn, and Mn under drought stress [54].

- In *Chrysanthemum morifolium* plant tissues, improvement in P and N contents were reported [55].

- In *Leymus chinensis*, an increased seedling weight by improving water content and intercellular CO2, P, and N contents was reported [56].

- *Glomus mosseae* and *Rhizophagus irregularis* showed improved heavy metal translocation in the shoot [57, 58].

4.1.8 Quality of soil

Mycorrhizal symbiosis can be further increased by agricultural practices like reduced tillage, low phosphorus fertilizer usage, and perennialized cropping systems [59, 60]. In the agroecosystems the quality of the soil and the productivity of the land can be enhanced by colonization of AM fungi. It enhances the constant masses, soil aggregate stability, rapidity of soil recovery, and significantly increases extra-radical hyphal mycelium in the soil. This is due to a soil protein known as glomalin, that is thought to be of AM fungal origin. Glomalin is responsible for improving soil aggregate water stability and for decreasing soil erosion [61].

4.1.9 Water stress tolerance

By physiological alteration of the above-ground organs and tissues, it enhances water stress tolerance, accumulation of dry matter and water moisture uptake, thereby improving plant tolerance against stresses like salinity and drought. Glomalin-related soil protein (GRSP) is maintain water content in soils exposed to different abiotic stresses and enhances the soil water holding capacity, which later on regulates water frequencies between soil and plants, thereby enhancing plant development [62–64].

4.1.10 Plant tolerance to stressful circumstances

Plant tolerance against various biotic and abiotic stressful circumstances like alkalinity and toxicity resulting from mining operations, heavy metals and mineral imbalance are reported to be increased by AM symbiosis. This is because of the communal nutrients' relocation from fungi to the plant, along with other related effects such as changes in their morpho-physiological traits [65–72].

4.1.11 Disease control

Apart from increasing the availability of macro and micronutrients, AM provides the plant with necessary strength to resist disease germs and unfavorable conditions. They also increase host tolerance to pathogen attack and compensate for the loss of root biomass or function caused by pathogens including Root-knot nematodes and fungi [73, 74]. The presence of AM fungi showed consistent reduction of disease symptoms for fungal pathogens such as *Phytophthora, Fusarium, Chalara, Pythium, Rhizoctonia, Sclerotium, Verticillium, Aphanomyces.* Several hypotheses have been put forwarded to explain the mechanisms of plant disease control by mycorrhizal fungi [75–77].

Some of them include creating a mechanical barrier for the pathogen penetration, thickening of cell wall through lignification and polysaccharide production that stops the entry of root pathogen, stimulation of the host roots to produce and accumulate sufficient concentration of metabolites like terpenes and phenols, imparting resistance to the host tissue against pathogen invasion, stimulating flavonolic wall infusions to prevent lesion formation by the pathogen, producing antifungal and antibacterial antibiotics, competing with the pathogens for the uptake of essential nutrients in the rhizosphere and at the roots surface, competitions in the roots and thus preventing the pathogen to get access to the roots. Harboring more actinomycetes antagonistic to root pathogen [78–81].

5. Conclusion and future prospect

Modern sustainable agriculture demands for a low-input and more nature-based system having role of soil loving microorganisms that are able to accelerate plant nutrition, health and soil, quality, also under stressful environments. All of these demands are being fulfilled by AM fungi. Its use in increasing food production is far and wide; therefore, is a better tool for modern sustainable agriculture particularly as biocontrol agent. Encouragement of AM as a tool for sustainable agriculture usage is of immense importance. Exploitation of AM for promoting a bio-healthy agriculture can significantly reduce the use of synthetic fertilizers and other chemicals resulting in agricultural improvement. Hence, using AM fungi as a biocontrol agent in modern sustainable agriculture, in terms of various parameters like reduction of damage caused by various pathogens, cost effectiveness, energy saving and also as an environment friendly, is a promising perspective for a sustainable agricultural system. The primary focus of future research should be on the identification of genes and gene products controlling the AMF mediated growth and development regulation under stressful cues. Identification of both host as well as AMF specific protein factors regulating symbiotic association and the major cellular and metabolic pathways under different environmental stresses can be hot areas for future research in this field.

Author details

Kavita Chahal[1*], Vaishali Gupta[1], Naveen Kumar Verma[1], Anand Chaurasia[2] and Babita Rana[3]

1 Department of Botany, Government College, Bichhua, Chhindwara, Madhya Pradesh, India

2 Department of Botany, Government College, Pawai, Panna, Madhya Pradesh, India

3 Department of Botany, Guru Nanak Khalsa College of Arts, Science and Commerce, Matunga, Mumbai, India

*Address all correspondence to: kavitachahal18@gmail.com

References

[1] Wheeler CT, Miller IM. Current and Potential Uses of Actinorhizal Plants in Europe, The Biology of Frankia and Actinorhizal Plants. 1990;7(11):365-389. Available: http://dx.doi.org/10.1016/b978-0-12-633210-0.50023-x

[2] Bethlenfalvay GJ., Schüepp H. Arbuscular mycorrhizas and agrosystem stability, Impact of Arbuscular Mycorrhizas on Sustainable Agriculture and Natural Ecosystems. 1994;9(15):117-131. Available: http://dx.doi.org/10.1007/978-3-0348-8504-1_10

[3] Robinson CH, Smith S E, Read DJ, Mycorrhizal Symbiosis, 2nd edn., The Journal of Ecology, 1997;85(6): 925. Available: http://dx.doi.org/10.2307/2960617

[4] Linderman RG, Vesicular-Arbuscular Mycorrhizae and Soil Microbial Interactions, Mycorrhizae in Sustainable Agriculture, ASA Special Publications. 2015;11(21): 45-70. Available: http://dx.doi.org/10.2134/asaspecpub54.c3

[5] Hepper C M. Isolation and Culture of VA Mycorrhizal (VAM) Fungi, VA Mycorrhiza 7(17): 95-112. Available: http://dx.doi.org/10.1201/9781351077514-5.

[6] Koske RE. Spores of VAM Fungi inside Spores of VAM Fungi, Mycologia 1984; 76(5): 853. Available: http://dx.doi.org/10.2307/3793141.

[7] Linderman RG. Vesicular-Arbuscular Mycorrhizal (VAM) Fungi, Plant Relationships Part B. 1997; 117-128. Available: http://dx.doi.org/10.1007/978-3-642-60647-2_7.

[8] Francis CA. Sustainable Agriculture Reviews, Sustainable Agriculture Reviews. 2015; 6(18):334. Available: http://dx.doi.org/10.1007/978-3-319-09132-7.

[9] Lal R. Sustainable Agriculture Reviews, Sustainable Agriculture Reviews. 2017; 8(11):67. Available: http://dx.doi.org/10.1007/978-3-319-58679-3.

[10] Kumm KV. Towards Sustainable Swedish Agriculture, Journal of Sustainable Agriculture. 2001;18(4): 27-37. Available: http://dx.doi.org/10.1300/j064v18n04_05.

[11] Zahedi H. Bioenergy and Sustainable Agriculture, Sustainable Agriculture Reviews, Sustainable Agriculture Reviews. 2018; 33:311-329. Available: http://dx.doi.org/10.1007/978-3-319-99076-7_11.

[12] Lal R. Soils and Sustainable Agriculture: A Review, Sustainable Agriculture. 2009: 15-23. Available: http://dx.doi.org/10.1007/978-90-481-2666-8_3.

[13] Benckiser G. Ants and Sustainable Agriculture, Sustainable Agriculture. 2011;2:15-26. Available: http://dx.doi.org/10.1007/978-94-007-0394-0_2.

[14] Suh J. Sustainable Agriculture in the Republic of Korea, Sustainable Agriculture Reviews, Sustainable Agriculture Reviews, 2019; 27:193-211. Available: http://dx.doi.org/10.1007/978-3-319-75190-0_7.

[15] Błaszkowski J. Mycorrhiza. 2002;12(6):317-317. Available: http://dx.doi.org/10.1007/s00572-002-0192-7.

[16] Albrechtova J. Fifth International Conference on Mycorrhiza. Granada, Spain, Mycorrhiza. 2006;16(3):227-227. Available: http://dx.doi.org/10.1007/s00572-006-0042-0.

[17] Vosatka M, Albrechtova J. Mycorrhiza for science and society—5th International Conference on Mycorrhiza (ICOM5), Mycorrhiza.

2006;17(2):157-158. Available: http://dx.doi.org/10.1007/s00572-006-0087-0.

[18] Pearson VG, Molina R. Mycorrhiza, Mycorrhiza, 2009; 19(2):67-67. Available: http://dx.doi.org/10.1007/s00572-008-0214-1.

[19] Habte M, Manjunath A. Categories of vesicular-arbuscular mycorrhizal dependency of host species, Mycorrhiza. 1991;1;(1):3-12. Available: http://dx.doi.org/10.1007/bf00205896.

[20] Janos DP. Vesicular-arbuscular mycorrhizae of epiphytes, Mycorrhiza. 1993; 4(1):1-4. Available: http://dx.doi.org/10.1007/bf00203242.

[21] Auge RM. Water relations, drought and vesicular-arbuscular mycorrhizal symbiosis, Mycorrhiza. 2001;11(1): 3-42. Available: http://dx.doi.org/10.1007/s005720100097.

[22] Nasr AA. Effects of vesicular-arbuscular mycorrhiza on *Tagetes erecta* and *Zinnia elegans*, Mycorrhiza. 1995;6(1):61-64. Available: http://dx.doi.org/10.1007/s005720050107.

[23] Kelly RM. The effects of vesicular arbuscular mycorrhizal fungi on the nutrition of sugarcane. 2017. Available: http://dx.doi.org/10.14264/uql.2017.345.

[24] Nedumpara MJ. Interactions of vesicular arbuscular mycorrhizal fungi, herbicides and crops. 2011. Available: http://dx.doi.org/10.31274/rtd-180813-10431.

[25] Miransari M. Plant, Mycorrhizal Fungi, and Bacterial Network, Plant signaling: Understanding the molecular crosstalk. 2013;315-325. Available: http://dx.doi.org/10.1007/978-81-322-1542-4_18.

[26] Raven JA. Why are mycorrhizal fungi and symbiotic nitrogen-fixing bacteria not genetically integrated into plants? Annals of Applied Biology. 2010;157(3): 381-391. Available: http://dx.doi.org/10.1111/j.1744-7348.2010.00435.x.

[27] Sanjuan J. Towards the minimal nitrogen-fixing symbiotic genome, Environmental Microbiology. 2016;18(8):2292-2294. Available: http://dx.doi.org/10.1111/1462-2920.13261.

[28] Thomas L, Rahman ZA. Genome-Wide Investigation on Symbiotic Nitrogen-Fixing Bacteria in Leguminous Plants, Plant Microbe Symbiosis. 2020; 7:55-73. Available: http://dx.doi.org/10.1007/978-3-030-36248-5_4.

[29] Sylvia DM. Distribution, Structure, and Function of External Hyphae of Vesicular-Arbuscular Mycorrhizal Fungi, Rhizosphere Dynamics.2019;8:144-167. Available: http://dx.doi.org/10.1201/9780429304798-6.

[30] Taylor LL. Review of manuscript gmd-2017-170 by He et al: Simulating ectomycorrhiza in boreal forests. 2017. Available: http://dx.doi.org/10.5194/gmd-2017-170-rc1.

[31] Wilson ID. Cotton Maturity and Plant Growth Regulators, Grow: Plant Health Exchange. 2018. Available: http://dx.doi.org/10.1094/grow-cot-12-18-172.

[32] Abdellatif L, Lokuruge P, Hamel C. Axenic growth of the arbuscular mycorrhizal fungus Rhizophagus irregularis and growth stimulation by coculture with plant growth-promoting rhizobacteria, Mycorrhiza. 2019;29(6):591-598. Available: http://dx.doi.org/10.1007/s00572-019-00924-z.

[33] Peng SHT, Yap CK, Arshad R, Chai EW. Bio-organic, Bio-chemical Fertilizers and N-Fixer (N-Bio Booster) Improve Paddy Yields in the Field Trials at Langkat in Medan, Indonesia. 2020. Available: http://dx.doi.org/10.20944/preprints202007.0584.v1.

[34] Kumar A, Singh JS. Microalgal bio-fertilizers, Handbook of Microalgae-Based Processes and Products. 2020: 445-463. Available: http://dx.doi.org/10.1016/b978-0-12-818536-0.00017-8.

[35] Dodd JC, Arias I, Koomen I, Hayman DS. The management of populations of vesicular-arbuscular mycorrhizal fungi in acid-infertile soils of a savanna ecosystem, Plant and Soil. 1990;122(2):241-247. Available: http://dx.doi.org/10.1007/bf02851981.

[36] Motta J. Effect of Glomus versiforme and Trichodema harzianum on growth and quality of Salvia miltiorrhiza, China Journal of Chinese Materia Medica. 2014. Available: http://dx.doi.org/10.4268/cjcmm20140906.

[37] Singh R, Behl RK, Singh KP, Jain P, Narula N. Performance and gene effects for wheat yield under inoculation of arbuscular mycorrhiza fungi and *Azotobacter chroococcum*, Plant, Soil and Environment. 2011;50(9):409-415. Available: http://dx.doi.org/10.17221/4052-pse.

[38] Posta K, Hong Duc N. Benefits of Arbuscular Mycorrhizal Fungi Application to Crop Production under Water Scarcity, Drought - Detection and Solutions, Gabrijel Ondrasek, IntechOpen, 2019. Available from: https://www.intechopen.com/books/drought-detection-and-solutions/benefits-of-arbuscular-mycorrhizal-fungi-application-to-crop-production-under-water-scarcity.

[39] Naderi NM, Alizadeh O, Nasr AH. Notice of Retraction: Some macro nutrients uptake optimizing by effect of Mycorrhizae fungi in water stress condition in sorghum plant, 2010. International Conference on Environmental Engineering and Applications. Available: http://dx.doi.org/10.1109/iceea.2010.5596119.

[40] Geneva M, Zehirov G, Djonova E, Kaloyanova N, Georgiev G, Stancheva I. The effect of inoculation of pea plants with mycorrhizal fungi and Rhizobium on nitrogen and phosphorus assimilation, Plant, Soil and Environment. 2011; 52(10):435-440. Available: http://dx.doi.org/10.17221/3463-pse.

[41] Qin L, Jiang H, Tian J, Zhao J, Liao H. Rhizobia enhance acquisition of phosphorus from different sources by soybean plants, Plant and Soil. 2011; 349(1-2):25-36. Available: http://dx.doi.org/10.1007/s11104-011-0947-z.

[42] Winkelmann G. A search for glomuferrin: a potential siderophore of arbuscular mycorrhizal fungi of the genus Glomus. 2017;30. Available: http://www.ncbi.nlm.nih.gov/pubmed/28616783.

[43] Labidi S, Jeddi FB, Tisserant B, Yousfi M, Sanaa M, Dalpé Y, et al. Field application of mycorrhizal bio-inoculants affects the mineral uptake of a forage legume (*Hedysarum coronarium* L.) on a highly calcareous soil. Mycorrhiza. 2015; 25:297-309. DOI: 10.1007/s00572-014-0609-0

[44] Charvat A. The mycorrhizal status of an emergent aquatic. 1999; 94: 191-197.

[45] Kobae Y, Tamura Y, Takai S, Banba M, Hata S. Localized expression of arbuscular mycorrhiza-inducible ammonium transporters in soybean. Plant & Cell Physiology. 2010; 51:1411-1415. DOI: 10.1093/pcp/pcq099.

[46] Hammer EC. Phosphorus and carbon availability regulate structural composition and complexity of AM fungal mycelium. 2014; 246: 443-451.

[47] Varma A. Mobilization of Micronutrients by Mycorrhizal Fungi. 2017; 9-26.

[48] Briccoli Bati C, Santilli E, Lombardo L. Effect of arbuscular mycorrhizal fungi on growth and on micronutrient and macronutrient uptake and allocation in olive plantlets growing under high total Mn levels. Mycorrhiza. 2015;25:97-108. DOI: 10.1007/s00572-014-0589-0

[49] Bedini S. Corrigendum to "Enhancing ecosystem services. 2014; 75: 314-315.

[50] Kobae Y. Investigation of Indigenous Arbuscular Mycorrhizal Performance Using a Lotus japonicus. Mycorrhizal Mutant. 2020; 95: 658.

[51] Sylvia DM. Aeroponic Culture of AM Fungi. 1999. 427-441.

[52] Rasmann S. Mycorrhizal Fungi Enhance Resistance to Herbivores. 2019; 93: 131.

[53] Ezawa T. Phosphorus metabolism and transport. 2019; 197-216.

[54] Cardoso IM, Kuyper TW. Mycorrhizas and tropical soil fertility. Agriculture, Ecosystems and Environment. 2006; 116:72-84. DOI: 10.1016/j.agee.2006.03.011.

[55] Franken P. Arbuscular mycorrhizal symbiosis. 2016; 217-238.

[56] Turnau K. Arbuscular Mycorrhiza. 2009; 87-111.

[57] Parray JA. Use of Mycorrhiza as Metal Tolerance Strategy. 2016; 57-68.

[58] Liu F. Arbuscular mycorrhiza improve growth. 2015; 262: 133-140.

[59] Angers D. Mycorrhiza and soil quality. 2004; 844: 353-353.

[60] Kaul HP. Arbuscular mycorrhiza enhances nutrient uptake. 2011; 5710: 465-470.

[61] Van Geel M, De Beenhouwer M, Lievens B, Honnay O. Crop-specific and singlespecies mycorrhizal inoculation is the best approach to improve crop growth in controlled environments. Agronomy for Sustainable Development. 2016; 36:37. DOI: 10.1007/s13593-016-0373-y.

[62] Kothe E. Modulation of ethanol stress tolerance by aldehyde dehydrogenase. 2011; 226: 471-484.

[63] Rahimzadeh S, Pirzad A. Pseudomonas and mycorrhizal fungi co-inoculation alter seed quality of flax under various water supply conditions. Industrial Crops and Products. 2019; 129:518-524. DOI: 10.1016/j.indcrop.2018.12.038

[64] Morte A., Lozano-Carrillo AC. 2010; 214: 247-253.

[65] Volpe V, Chitarra W, Cascone P, Volpe MG, Bartolini P, Moneti G, et al. The association with two different arbuscular mycorrhizal fungi differently affects water stress tolerance in tomato. Frontiers in Plant Science. 2018;9:1480. DOI: 10.3389/fpls.2018.01480

[66] Zhang F., 2002 Improved tolerance of maize plants to salt stress by arbuscular mycorrhiza is related to higher accumulation of soluble sugars. 124: 185-190.

[67] Hu W, Zhang H, Zhang X, Chen H, Tang M. Characterization of six PHT1 members in *Lycium barbarum* and their response to arbuscular mycorrhiza and water stress. Tree Physiology. 2017;37:351-366. DOI: 10.1093/treephys/tpw125

[68] Miransari M. Recent Advances on Mycorrhizal Fungi, Fungal Biology. 2016; 63-79.

[69] Giri B. Arbuscular Mycorrhiza: Approaches for Abiotic Stress Tolerance. 2012; 359-401.

[70] Martin-Robles N, Lehmann A, Seco E, Aroca R, Rillig MC, Milla R. Impacts of domestication on the arbuscular mycorrhizal symbiosis of 27 crop species. New Phytologist. 2018; 218:322-334. DOI: 10.1111/nph.14962 Alena Andrejiová, 2018 Stress Tolerance of Vegetables: A Review. 212: 30-35.

[71] Azmat R and Moin S. The remediation of drought stress under AM inoculation through proline chemical transformation action. 2019; 193:6.

[72] Calvet C. The contribution of arbuscular mycorrhizal fungi to the control of soil-borne plant pathogens. 2005; 187-197.

[73] Watts-Williams SJ, Cavagnaro TR, Tyerman SD. Variable effects of arbuscular mycorrhizal fungal inoculation on physiological and molecular measures of root and stomatal conductance of diverse *Medicago truncatula* accessions. Plant, Cell & Environment. 2019;42:285-294. DOI: 10.1111/pce.13369

[74] Duc NH, Posta K. Mycorrhizainduced alleviation of plant disease caused by Clavibacter michiganensis subsp. michiganensis and role of ethylene in mycorrhiza-induced resistance in tomato. Acta Biologica Hungarica. 2018;69(2):170-181. DOI: 10.1556/018.69.2018.2.6.

[75] Berta G. Mycorrhiza-induced differential response to a yellows disease. 2002; 124: 191-198.

[76] M Singh. Biological control of Fusarium wilt of tomato by arbuscular mycorrhizal fungi with intercropping. 101: 1-9.

[77] Goltapeh EM, Danesh YR, Prasad R, Varma A. Mycorrhizal fungi: What we know and what should we know? In: Varma A, editor. Mycorrhiza. 3rd ed. Berlin Heidelberg: Springer-Verlag; 2008. pp. 3-27

[78] Giri B. Arbuscular Mycorrhiza Mediated Control of Plant Pathogens. 2017; 131-160.

[79] Mukerji KG. Allelochemicals: Biological Control of Plant Pathogens. 2006; 181-192.

[80] Berruti A, Borriello R, Lumini E, Scariot V, Bianciotto V, Balestrini R. Application of laser microdissection to identify the mycorrhizal fungi that establish arbuscules inside root cells. Frontiers in Plant Science. 2013; 4:135. DOI: 10.3389/fpls.2013.00135

[81] Borriello R, Lumini E, Girlanda M, Bonfante P, Bianciotto V. Effects of different management practices on arbuscular mycorrhizal fungal Drought - Detection and Solutions 26 diversity in maize fields by a molecular approach. Biology and Fertility of Soils. 2012;48:911-922. DOI: 10.1007/s00374-012-0683-4

Chapter 3

Advantages of Arbuscular Mycorrhizal Fungi (AMF) Production for the Profitability of Agriculture and Biofertilizer Industry

Santhi Sudha Samuel and Aranganathan Veeramani

Abstract

Decades of ill-agricultural practices associated with emerging risks of climatic changes have been degrading the ecosystem with immense stress on the soil health, crop productivity. Arbuscular mycorrhiza (AM) form advantageous symbiosis between plant roots and specialized soil fungi that is rampant in natural habitats. Studies show that the elevated AMF indicated good soil health, high crop turnouts benefiting the Agriculture and other industries. AMF dependent on plants for sugars, while offering benefits like intact binding of soil particles, biomass increase, improvement of water-holding capacity, replacement of harmful chemicals, increased intake of phosphorous, zinc and other nutrients, drought and salinity tolerance, carbon sequestering in soil and protection from nematodes and other predatory insects. AMF are best candidates as bio-fertilizers and this review will explore their beneficial interconnections.

Keywords: arbuscular mycorrhizal fungi (AMF), bio-fertilizers, phosphorous, drought and salinity tolerance, nematodes

1. Introduction

In past decades, there are escalating events of abiotic stress like drought, low or high-salinity, soils contaminated with heavy metals, extremely high or low temperatures and other extreme calamities of climate such as hurricanes, tornadoes have always been detrimental to agriculture and industries. Added to the list, increasing effects of the unseasonal climatic changes such as earthquakes, tsunamis have immensely damaged our lands. On top of everything else, agricultural malpractices like poorly managed animal feeding operations, overgrazing, plowing, tilling, excessive fertilizer usage, genetically modified crops, deforestation, excessive irrigation, pesticides, phosphorous mining, poor agricultural waste management, increasing soil pathogens have severely crippled the agricultural yields [1]. The world populace is projected to be 9.7 billion in 2050 and approximately 11 billion around 2100 according to UN population division and thus the global need for increased food sources by agricultural production must be definitely promoted to keep up with the population. Many improvements must come up for better agricultural quality and productivity shortcomings with insightful management.

Figure 1.
Mycorrhizal fungal association in roots of legumes (photo courtesy: Corsi and Muminjanov [2]).

The Food and Agriculture Organization (FAO) United Nations reveals major challenges to sustainable intensification of land and agricultural practices such as land degradation due to tillage erosion, soil compaction, overgrazing, nutrient mining, overuse of mineral fertilizers and herbicides, inefficient irrigation practices, ignorant crop management practices and other malpractices [2] that needs immediate attention. The FAO Agricultural Development Economics Division voice out for crucial remedial measures on agricultural practices as the population growth and global food demand towards 2030/50 seems worrisome [3].

Studies show that AMF associated agricultural practices can offer major relief by sustainable and beneficial consequences for both agricultural and natural ecosystems through its association with the plants and soil. This chapter presents the symbiotic interconnections of AMF for the advantageousness in agriculture and industry are highlighted (**Figure 1**).

2. AMF background and evolvement

The fungus is a eukaryotic organism that belonged to the plant kingdom at the beginning; nonetheless, as the studies unfolded its unique features, scientists realized that fungi were much alike animals than plants by exhibiting features like cell membrane-bound organelles, distinct nuclei, and lack of chlorophyll in their cells. Fungi have an exclusive life cycle of having principal modes of vegetative growth, nutrient intake unlike animals that enabled it to have a distinguished identity. The kingdom of fungi has about 1.5 million known species covering the yeasts, rusts, smuts, mildews, molds, and mushrooms. Some organisms like slime molds and oomycetes (water molds) show many fungi-like features and are included in the kingdom of Chromista. Fungi are widely distributed on earth and are mostly free-living, some form parasitic or symbiotic relationships with plants or animals. Many fungi are of great environmental and medical importance and its study is termed as Mycology [4].

A typical fungus has minute filamentous cytoplasmic morphologies bounded by a plasma membrane and cell wall known as hyphae. Their cell walls are made up of flexible polysaccharides called chitin, resembling the exoskeleton of insects. The hyphae have many auxiliary cell walls, known as cross-walls or septa, typically perforated with pores that are large enough for ribosomes, mitochondria, and nuclei to flow. The hyphae branch extensively as they mature, forming complex multicellular structures known as mycelium. A mycelium gives heterotrophic nourishment as they feed on organic food sources, digest them externally before absorption by secreting valuable enzymes into its surroundings. These mycelia can spread to vast areas that

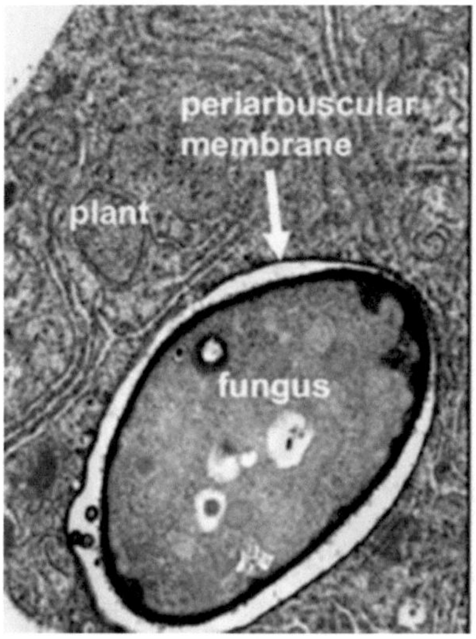

Figure 2.
Transmission Electron micrographic view of a colonized host cell with an arbuscular branch (fungus), surrounded by the peri arbuscular membrane (photo courtesy: Chen et al. [8]).

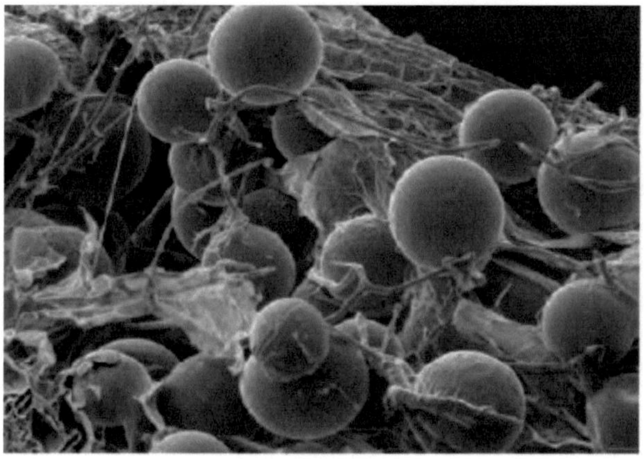

Figure 3.
Scanning electron micrograph of mycorrhizal hyphae and spores (photo curtesy: Mycorrhizal applications @ GPNMAG.COM 2018).

serve as a phenomenal symbiotic benefit to the plant root system, giving it a unique chance to obtain phosphate and other minerals far off the nutriment depleted zones, while the fungal takes sugars from the plant. Symbiotic association between a fungus and the roots of a vascular plant is often termed as Mycorrhiza or root fungi [5].

The mycorrhization of a plant root that develops as a cover surrounding the roots is termed mantle, from which the hyphae grow, and this is Ectomycorrhiza. On the other hand, Endomycorrhiza is when the mycelium is implanted within the root tissue and these are also termed as the arbuscular mycorrhizae (AM) or arbuscular mycorrhizal fungi (AMF).

More than 80% of the terrestrial plants show symbiosis with AMF and they belong to Phylum Glomeromycota. AMFs are obligate biotrophs, absorb photosynthetic byproducts and lipids in the plant symbiotic connections [6]. AM are expected to have originated approximately 480 million years ago, the fossilized fungal hyphae and spores from the Ordovician of Wisconsin (USA) bear a strong resemblance with the current AMF (Glomales, Zygomycetes). These fossils suggested that Glomales-like fungi existed in the bryophytic vegetation. Later many reports supported that AMF was essential constituents in predominant land plants in most taxa in all the ecological niches for ages. The Glomeromycota are a distinctive obligate biotrophic fungi that majorly comprises AMF related in symbiosis with many Embryophytes [7]. Nevertheless, certain mycoheterotroph plant species on AMF symbiosis turned in to obligate parasites having completely lost plastid genetic apparatus, photosynthesis genes with secondary functions, NADH dehydrogenase-like genes and photosynthesis genes. As another diversity was seen in plant taxa such as Brassicacea (or Cruciferae) and Chenopodiaceae, where they have shown asymbiotic interaction with AMF and these plants developed other strategies for their nutritional requirements (**Figures 2** and **3**) [9].

3. AMF as biofertilizers

Many active or dormant strains of bacteria or fungi or in combinations are used diametrically or collaterally to activate the rhizo-microbiome and trigger the nutrient supply from soil to plants that would ultimately result in enhancement of crop yields. These microbial strains are broadly termed as biofertilizers, bio-inoculants, agricultural inoculants, soil inoculants, or microbial inoculants. These bio-inoculants with unique merits are highly encouraged globally and are earning prominence in modern agricultural customs, practices and maneuvers contrasting to other conventional or synthetic pesticides and fertilizers. These biofertilizers are safe to handle, are required only in small quantities as they capable of fast replication, their action can be leveraged or optimized based on their incumbencies, decompose quickly with negligible ill effects to the environment and show lower resistance to host plant and infective organisms [10].

Plant growth promoting and disease suppressing microbial inoculants such as *Azospirillum, Bacillus, Pseudomonas, Rhizobium, Serratia, Stenotrophomonas,* and *Streptomyces, Ampelomyces, Coniothyrium, Glomus,* and *Trichoderma are* extensively examined and analyzed for their mechanism of action and regulatory gear. Even though multiple categories of biofertilizers are at hand, the AMF is reported of manifold advantages and graded high for soil health and crop productivity [11]. Studies have shown comparative progress where plants get better nourishment with greater AMF colonization than that of the non-mycorrhizal plants even with conditions like mineral deficiency and abiotic stress. AMF establishes symbiotic interconnection with many different types of Plants or the other way round, that leads to the formation of common mycorrhizal networks (CMNs) and such mycorrhizal interactions are exceptionally valuable for healthier plant growth and yield in most of the scenarios [12].

Certain AMF show specific combinatorial benefits with specific types of plant species from all types of geographical locations that can result in positive mycorrhizal growth response (MGR) and this is a progressive mutual adaptation that did not correlate with phylogenetic lineage patterns relevant to variant species [13]. Contrarily, some studies though agree with the functional specialization of AMF, proved that such incidence is a flexible phenomenon where plant species are required to show compatibility with at least a few AMF. This scenario with minimal host specificity and broad functional specialization encourages vast biodiversity and productivity in plant communities [14].

Rampant use of inorganic fertilizers, herbicides, and fungicides are causing multiple injurious health risks to every living organism by hazardous impacts on the quality of food, soil, air, and water systems [15]. Over the years, many investigations have proven the efficacy of AMF for best agricultural production compared to other synthetic or chemical fertilizers under the prevalent stressful conditions, as addressed below.

4. Drought tolerance

Drought is the scarcity of water in soil for prolonged durations affects plant growth. It has severe implications on the entire plant biorhythm and growth at every notch. Deficit water supply to roots causes oxidative stress due to anomalies in transpiration [16], affects enzyme activity, ion uptake, and nutrient assimilation [17]. Many investigations have evidenced that AMF can allay drought stress in varied crop like wheat, barley, maize, soybean, strawberry, and onion [18]. This remarkable tolerance is reasoned essentially due to the extra-radical hyphae of AMF that has the capability of vast area spread [19]. Further, the osmotic adjustment, stomatal regulation, enhanced proline, and glutathione level are exhibited to have augmented root efficiency, leaf area index, and biomass under the instant drought conditions and against severe environmental conditions. Reports have demonstrated that the enhancement in growth and photosynthesis in C3 (*Leymus chinensis*) and C4 (*Hemarthria altissima*) plant species through up-regulation of antioxidant system by AMF symbiosis (**Figure 4**) [20].

Figure 4.
2017 California spring trials. Coreopsis plants (image 1) inoculated with AMF (left) showed better tolerance than plants without AMF under same drought stress. Coreopsis plants (image 2) treated with AMF showed improved top growth and root system development (left) than that without AMF inoculation (right) (photo courtesy: Mycorrhizal applications @ GPNMAG.COM 2018).

5. Salinity stress alleviation

Soil salinization is an aggravating issue threatening global food safety as it suppresses the plant development leading to reduced crop harvest (due to enormous formation of reactive oxygen species (ROS)) [21]. Many research reports showed the efficiency of AMF to enhance growth and crop yield under salinity stress. AMF association triggered the synthesis of plant hormones such as jasmonic acid and salicylic acid, and inorganic nutrients (P, Ca^{2+}, N, Mg^{2+}, and K^+) under salt stress conditions [22]. Some mycorrhizal associated plants showed increased amount of biomass, proline, N_2, and remarkable alteration in ionic uptake. AMF inoculation showed better levels of key growth regulators such as cytokinin, polyamine and strigolactone concentrations, suppressing lipid membrane peroxidation and regulation of the osmoregulation [23].

6. Resilience to extreme temperatures

Extreme temperatures such as Heat stress and Cold stress are prevalent challenges faced by plants globally. Heat stresses reduce seed germination, growth rate and biomass, and cause wilting or burning of leaves and reproductive organs, and which leads to senescence of leaves, damage and discoloration of fruit, reduction in yield, cell death, and enhanced oxidative stress [16]. Mycorrhizal plants showed encouraging growth under the conditions of high temperature [24]. AMF supports plants in cold stress and helps plant development [25] as they can retain moisture in the plant [26], increase plant secondary metabolites boosting immune system, and improve protein content to ameliorate cold stress [27].

7. Minimization of heavy metal toxicity

Accumulation of heavy metals in food crops, fruits, vegetables, and soils are very hazardous [28]. Plants grown on soils with excessive Cd and Zn exhibit considerable suppression in shoot, root growth, leaf chlorosis, and even death [29]. AMF associations have shown fortified growth and crop yield under aluminum stress and other metals [30]. Heavy metals are immobilized in the internal or external surface of fungal hyphae and will be stored in their vacuoles or may chelate with some other substances in the cytoplasm, minimizing the toxicity effects [31]. Mycelia of various AMF have a high cation-exchange capacity and absorption of metals [32] and they enhance the plant biomass, that uptake important immovable nutrients like Cu, Zn, and P further nullifies the metal toxicity [33].

8. Oxidative stress

Exposure of plants to drought, salinity, heat and cold stress, other harmful conditions causes oxidative stress, an enhanced production of reactive oxygen species (ROS), which can be highly injurious to plants [34]. Some of the enzymes including superoxide dismutase (SOD), catalase (CAT), peroxidase (POD), and glutathione reductase (GR) prevent the production of ROS [35]. Mycorrhizal plants have proven to successfully overcome oxidative stress by improved biomass production, leaf water relations and stomatal conductance [36], other amplifying operations include improved photosynthetic rate, uptake and accumulation of minerals, assemblage of osmo-protectants, up-regulation of antioxidant enzyme activity, and change in the rhizosphere ecosystem [37]. Studies have shown the improved nutritional status of AMF plants under osmotic stress conditions resulting from deficit irrigation or salinity. Some of the substantial variations were observed in the charectorstics of phytohormones, absorption of minerals, compilation of osmolytes and secondary metabolites, and in the antioxidant execution systems. These impacts are presumed to have enhanced the nutraceutical value of yield in crops mounting to immense agronomic accomplishments [38].

9. Biotic stress

Plants encounter biotic stresses caused by pathogenic fungi, viruses, bacteria, nematodes, insects, etc. that cause diseases, infections and affect crop productivity. AMF association is known for potential biocontrol mechanisms such as antibiotic production, competitive interaction strategies among the rhizo-microbiome and pathogens, mycoparasitism, and inducing genetic expression changes able to induce systemic resistance inside the host plant [39]. AMF is reported to have biocontrol capacity over powdery mildews [40] and the nonpathogenic and saprotrophic species of Rhizoctonia, Fusarium, and Trichoderma have been utilized to reduce damage caused by genetically and phenotypically similar pathogenic fungi. They are known to control soil-borne and plant diseases and studies reported that increased plant growth was seen in associated with strains of *Trichoderma, Glomus intraradices, Glomus mosseae*, and other plant growth-promoting microorganisms due to systemic resistance against plant pathogens by upregulating specific genes in the host plant [41]. There is a growing body of evidence on multifunctional prospects of AMF as efficient biocontrol agents for augmented plant productivity by enhancing crop nutrition. Many AMF act as broad biopesticides or selective agents such as mycoinsecticides, mycoacaricides, myconematicides and others [42].

10. Minerals and phosphorous transport

AMF improves plant nutrition and helps them to cope with changing environments. Plants use inorganic phosphate (Pi) and it is the most important limiting factor for its growth. Since soluble Pi levels are low in the soil, the symbiotic interconnection with AMF will efficiently supply the needful Pi and other mineral nutrients in exchange for carbohydrates [43]. AMF expresses proteins to transfer inorganic phosphate (Pi) from the soil to colonized roots through symbiotic interfaces [44]. AMF compatible plants have two Pi uptake pathways that have different sets of phosphate transporters: a direct up take pathway through the epidermis and root hairs, and a symbiotic uptake pathway for the Pi provided by the fungus [45]. In Addition, AMF shows extraordinary symbiotic Pi uptake, by boosting the plant mineral nutrient acquisition even with low-nutrient supply [46].

11. Crop yield enhancement

AMF can improve the nutrient status, quality, and yield of the crops, AMF-colonized crops show increased levels of secondary metabolites with antioxidant [47] and enhancement of dietary quality of crops with carotenoids and volatile compounds were observed [48]. Mycorrhizal symbiosis enhances the accumulation

Figure 5.
Effect of mycorrhizal treatment on corn (on the left) with their control (on the right). Ohio, USA, 2019 (photo courtesy: Groundwork BioAg).

of anthocyanins, chlorophyll, carotenoids, total soluble phenolics, tocopherols, sugars, organic acids, vitamin C, flavonoids, and mineral nutrients [49] and enhanced the biosynthesis of phytochemicals in edible plants (**Figure 5**) [50].

12. Soil erosion and nutrient leaching

AMF helps to successfully overcome soil erosion and nutrient leaching in natural as well as in agricultural lands [51]. AMF mycelium is highly ramified and creates a three-dimensional matrix that enmeshes and crosslinks soil particles without compacting the soil with soil glycoprotein, glomalin for stabilization of soil aggregates [52]. Glomalin and glomalin-related soil proteins (GRSPs) account for a vital fraction of total organic soil carbon (2–5%), and for sequestration of carbon in the soil [53]. The hyphal network of AMF, and their promoting effects on plant growth and root system development, protect the soil from erosion by wind and water, promotes water retention capacity and nutrient supply [54].

Nutrient leaching is a serious risk as it results in soil infertility and pollution of groundwater and surface water (rivers, lakes). Agricultural lands are disturbed by malpractices like extensive plowing and tilling, and receive large amounts of fertilizer with N, P and K. These can get washed out from the soil due to the lack of retention systems leading to undesirable consequences [55]. AMF reduces nutrient leaching from the soil [56] by operating at different levels, such as improving soil structure, nutrient sequestration to the micro and macro-soil aggregates, uptake of nutrients from the soil solution and reviving its retention capacity [57].

13. Crop care and horticulture

Cultivation of a garden, orchard, or nursery of flowers, fruits, vegetables, or ornamental plants with AMF associations has received notable appreciation. Other extended aspects of horticulture include plant conservation, landscape restoration, soil management, landscape and garden design, construction, maintenance, and arboriculture [58]. The AMF attributes such as upregulation of Pi, nutrients, survival rate, plantlet micropropagation, crop uniformity, enhanced fruit production with high nutrient values, resistance to biotic and abiotic stress are attracting botanists, horticulturists, and other scientists (**Figures 6** and 7) [60].

Figure 6.
Outcome of AMF inoculation on enhanced root structure and size of Banana saplings; control compared with the test saplings T_{1-4} (photo courtesy: Emara [59]).

Figure 7.
Geranium plants grown on commercial nutrient regime (left) and that grown with the same nutrient supply with additional mycorrhizal fungi (right) which shows better nutrient and water uptake and use efficiency (photo curtesy: Mycorrhizal applications @ GPNMAG.COM 2018).

14. Potential use in reforestation, landscaping, bioremediation and revegetation

Mycorrhizal associations are extensively utilized for reforestation programs, the ectomycorrhizal fungi are employed to produce containerized seedlings and AMF are raised with plantlets to survive transplantation shock [61]. Such seedlings may have better survival after planting in tropical settings or another natural environment with varied climate conditions [62].

Landscaping is an evolving industry financed by enthusiastic customers from private corporations, businesses, private homes and government agencies for esthetic highway and road maintenance, seeking low-cost and natural solutions. AMF association has encouraged the best native planting and reclamation practices with appealing sports fields, road medians, golf courses, public and private parks, and gardens [60].

Bioremediation and revegetation are a scenario that promotes plant growth in contaminated soils and AMF has acclaimed its potential in this regard also [63]. Decades of agricultural and industrial malpractices, volcanic ash, mine spoils, waste deposits and other anthropogenically polluted sites are filled with organic compounds or heavy metals. AMF has mineral-scavenging capacities with two kinds of strategies, they accumulate and sequester toxic metal ions, or they deliver to plants in the form of essential mineral nutrients such as Cu and Zn [64].

Figure 8.
Improved land scaping with AMF (photo courtesy: AMF lawns @ AMFLawns. Landscape Company).

Bioremediation and phytoremediation by mycorrhizal inoculants are an emerging frontier and needs attention (**Figure 8**).

15. Conclusion

AMF association with plants amplifies its growth and harvest with fortified nutrients, parallelly resists environmental stress and defends from infections. Furthermore, revamps soil quality, texture, and water retention capabilities in both the agricultural and industrial sectors. With all the fundamental features, they are proving to be significant in both academic and commercial arenas. Novel cost-effective techniques are required to check AMF quality control, social media, computer-based technologies can help to promote mycorrhizae application on agricultural field. The awareness presentations through social media will be a best method to reach the importance of AMF to organic farmers. Antimicrobial resistance is an evolved hazard in the modern world and replacing chemical fertilizers, synthetic pesticides, fertilizers, and other microbicides with biofertilizers is predominantly essential.

Author details

Santhi Sudha Samuel and Aranganathan Veeramani*
Department of Biochemistry, Jain (Deemed to-be) University,
Bengaluru, Karnataka, India

*Address all correspondence to: v.aranganathan@jainuniversity.ac.in

References

[1] Begum N, Qin C, Ahanger MA, Raza S, Khan MI, Ashraf M, Ahmed N, Zhang L. Role of Arbuscular Mycorrhizal Fungi in Plant Growth Regulation: Implications in Abiotic Stress Tolerance. Front. Plant Sci. 2019;10:1068. DOI: 10.3389/fpls.2019.01068.

[2] Corsi S, Muminjanov H. Conservation Agriculture: Training guide for extension agents and farmers in Eastern Europe and Central Asia. Rome. FAO. 2019.

[3] Alexandratos N, Bruinsma J. World agriculture towards 2030/2050: the 2012 revision. Rome, FAO. 2012;ESA Working paper No. 12-03.

[4] Bongomin F, Gago S, Oladele RO, Denning DW. Global and Multi-National Prevalence of Fungal Diseases-Estimate Precision. Journal of fungi (Basel, Switzerland). 2017;3(4), 57. DOI: 10.3390/jof30407.

[5] Wu B, Hussain M, Zhang W, Stadler M, Liu X, Xiang M. Current insights into fungal species diversity and perspective on naming the environmental DNA sequences of fungi. Mycology. 2019;10:3, 127-140. DOI:10.1080/21501203.2019.1614106.

[6] Jiang YN, Wang WX, Xie QJ, Liu N, Liu LX, Wang DP, Zhang X, Yang C, Chen X, Tang D, Wang E. Plants transfer lipids to sustain colonization by mutualistic mycorrhizal and parasitic fungi. Science. 2017;356, 1172-1175. DOI: 10.1126/science.aam9970.

[7] Delaux PM. Comparative phylogenomics of symbiotic associations. New Phytol. 2017;213, 89-94. DOI: 10.1111/nph.14161.

[8] Chen M, Arato M, Borghi L, Nouri E, Reinhardt D. Beneficial Services of Arbuscular Mycorrhizal Fungi – From Ecology to Application. Front. Plant Sci. 2018;9:1270. DOI: 10.3389/fpls.2018.01270

[9] Brundrett M. Diversity and classification of mycorrhizal associations. Biol. Rev. 2004;79, 473-495. DOI: 10.1017/S1464793103006316.

[10] Berg G. Plant-microbe interactions promoting plant growth and health: perspectives for controlled use of microorganisms in agriculture. Appl Microbiol Biotechnol. 2009;84(1):11-18. DOI: 10.1007/s00253-009-2092-7.

[11] Ortas I. The effect of mycorrhizal fungal inoculation on plant yield, nutrient uptake and inoculation effectiveness under long-term field conditions. Field Crops Res. 2012;125, 35-48. DOI: 10.1016/j.fcr.2011.08.005.

[12] Jakobsen I, Hammer EC. "Nutrient dynamics in arbuscular mycorrhizal networks," in Mycorrhizal Networks, ed. T. R. Horton (Dordrecht: Springer). 2018;91-131. DOI: 10.1007/978-94-017-7395-9_4.

[13] Kiers ET, Denison RF. Sanctions, cooperation, and the stability of plant-rhizosphere mutualisms. Annu. Rev. Ecol. Evol. Syst. 2008;39, 215-236. DOI: 10.1146/annurev.ecolsys.39.110707.173423.

[14] van der Heijden MGA, Klironomos JN, Ursic M, Moutoglis P, Streitwolf-Engel R, Boller T, Wiemken A, Sanders IR. Mycorrhizal fungal diversity determines plant biodiversity, ecosystem variability and productivity. Nature. 1998;396, 69-72. DOI: 10.1038/23932.

[15] Yang S, Li F, Malhi SS, Wang P, Dongrang S, Wang J. Long term fertilization effects on crop yield and nitrate nitrogen accumulation in

soil in Northwestern China. Agron. J. 2004;96, 1039-1049. DOI: 10.2134/agronj2004.1039.

[16] Hasanuzzaman M, Gill SS, Fujita M. "Physiological role of nitric oxide in plants grown under adverse environmental conditions," in Plant acclimation to environmental stress. Eds. N. Tuteja and S. S. Gill (NY: Springer Science+Business Media). 2013;269-322. DOI: 10.1007/978-1-4614-5001-6_11.

[17] Ahanger MA, Tittal M, Mir RA, Agarwal RM. Alleviation of water and osmotic stress-induced changes in nitrogen metabolizing enzymes in Triticum aestivum L. cultivars by potassium. Protoplasma. 2017;254 (5), 1953-1963. DOI: 10.1007/s00709-017-1086-z.

[18] Moradtalab N, Roghieh H, Nasser A, Tobias EH, Günter N. Silicon and the association with an arbuscular-mycorrhizal fungus (Rhizophagus clarus) mitigate the adverse effects of drought stress on strawberry. Agronomy. 2019;9, 41. DOI: 10.3390/agronomy9010041.

[19] Zhang X, Li W, Fang M, Jixian Y, Meng S. Effects of arbuscular mycorrhizal fungi inoculation on carbon and nitrogen distribution and grain yield and nutritional quality in rice (Oryza sativa L.). J. Sci. Food Agric. 2016;97, 2919-2925. DOI: 10.1002/jsfa.8129.

[20] Li J, Meng B, Chai H, Yang X, Song W, Li S, Lu A, Sang T, Sun W. Arbuscular mycorrhizal fungi alleviate drought stress in C3 (Leymus chinensis) and C4 (Hemarthria altissima) grasses via altering antioxidant enzyme activities and photosynthesis. Front. Plant Sci. 2019;10, 499. DOI: 10.3389/fpls.2019.00499.

[21] Ahanger MA, Alyemeni MN, Wijaya L, Alamri SA, Alam P, Ashraf M, Ahmed P. Potential of exogenously sourced kinetin in protecting Solanum lycopersicum from NaCl-induced oxidative stress through up-regulation of the antioxidant system, ascorbate–glutathione cycle and glyoxalase system. PLoS One 2018;13 (9), e0202–e0175. DOI: 10.1371/journal.pone.0202175.

[22] Hashem A, Alqarawi AA, Radhakrishnan R, Al-Arjani AF, Aldehaish HA, Egamberdieva D, Allah EFA. Arbuscular mycorrhizal fungi regulate the oxidative system, hormones and ionic equilibrium to trigger salt stress tolerance in Cucumis sativus L. Saudi J. Biol. Sci. 2018;25 (6), 1102-1114. DOI: 10.1016/j.sjbs.2018.03.009.

[23] Santander C, Sanhueza M, Olave J, Borie F, Valentine C, Cornejo P. Arbuscular mycorrhizal colonization promotes the tolerance to salt stress in lettuce plants through an efficient modification of ionic balance. J. Soil Sci. Plant Nutr. 2019;19 (2), 321-331. DOI: 10.1007/s42729-019-00032-z.

[24] Maya MA, Matsubara Y. Influence of arbuscular mycorrhiza on the growth and antioxidative activity in Cyclamen under heat stress. Mycorrhiza. 2013;23(5), 381-390. DOI: 10.1007/s00572-013-0477-z.

[25] Birhane E, Sterck F, Fetene M, Bongers F, Kuyper T. Arbuscular mycorrhizal fungi enhance photosynthesis, water use efficiency, and growth of frankincense seedlings under pulsed water availability conditions. Oecologia. 2012;169, 895-904. DOI: 10.1007/s00442-012-2258-3.

[26] Zhu XC, Song FB, Xu HW. Arbuscular mycorrhizae improve low temperature stress in maize via alterations in host water status and photosynthesis. Plant Soil. 2010;331, 129-137. DOI: 10.1007/s11104-009-0239-z.

[27] Abdel Latef AA, Chaoxing H. Arbuscular mycorrhizal influence on growth, photosynthetic pigments, osmotic adjustment and oxidative stress in tomato plants subjected to low temperature stress. Acta Physiol. Plant. 2011;33,1217-1225. DOI: 10.1007/s11738-010-0650-3.

[28] Yousaf B, Liu G, Wang R, Imtiaz M, Zia-ur-Rehman M, Munir MAM, Niu Z. Bioavailability evaluation, uptake of heavy metals and potential health risks via dietary exposure in urban-industrial areas. Environ. Sci. Pollut. Res. 2016;23, 22443-22453. DOI: 10.1007/s11356-016-7449-8.

[29] Moghadam HRT. Application of super absorbent polymer and ascorbic acid to mitigate deleterious effects of cadmium in wheat. Pesqui. Agropecu. Trop. 2016;6 (1), 9-18. DOI: 10.1590/1983-40632016v4638946.

[30] Aguilera P, Pablo C, Fernando B, Fritz O. Diversity of arbuscular mycorrhizal fungi associated with Triticum aestivum L. plants growing in an andosol with high aluminum level. Agri. Eco. Environ. 2014;186, 178-184. DOI: 10.1016/j.agee.2014.01.029.

[31] Punamiya P, Datta R, Sarkar D, Barber S, Patel M, Da P. Symbiotic role of Glomus mosseae in phytoextraction of lead in vetiver grass Chrysopogon zizanioides L. J. Hazard. Mater. 2010;177, 465-474. DOI: 10.1016/j.jhazmat.2009.12.056.

[32] Takács T, Vörös I. Effect of metal non-adapted arbuscular mycorrhizal fungi on Cd, Ni and Zn uptake by ryegrass. Acta Agron. Hung. 2003;51, 347-354.

[33] Miransari M. "Arbuscular mycorrhizal fungi and heavy metal tolerance in plants," in Arbuscular mycorrhizas and stress tolerance of plants. Ed. Q. S.Wu (Singapore: Springer Nature). 2017;174-161. DOI: 10.1007/978-3-319-68867-1_4.

[34] Bauddh K, Singh RP. Growth: tolerance efficiency and phytoremediation potential of Ricinus communis (L.) and Brassica juncea (L.) in salinity and drought affected cadmium contaminated soil. Ecotoxicol. Environ. Saf. 2012;85, 13-22. DOI: 10.1016/j.ecoenv.2012.08.019.

[35] Ahanger MA, Agarwal RM. Potassium up-regulates antioxidant metabolism and alleviates growth inhibition under water and osmotic stress in wheat (Triticum aestivum L.). Protoplasma. 2017;254 (4), 1471-1486. DOI: 10.1007/s00709-016-1037-0.

[36] Duc NH, Csintalan Z, Posta K. Arbuscular mycorrhizal fungi mitigate negative effects of combined drought and heat stress on tomato plants. Plant Physiol. Biochem. 2018;132, 297-307. DOI: 10.1016/j.plaphy.2018.09.011.

[37] Calvo-Polanco M, Sanchez-Romera B, Aroca R, Asins MJ, Declerck S, Dodd IC, Martinez-Andujar C, Albacete A, Lozano JMR. Exploring the use of recombinant inbred lines in combination with beneficial microbial inoculants (AM fungus and PGPR) to improve drought stress tolerance in tomato. Environ. Exp. Bot. 2016;131, 47-57. DOI: 10.1016/j.envexpbot.2016.06.015.

[38] Auge RM, Toler HD, Saxton AM. Arbuscular mycorrhizal symbiosis and osmotic adjustment in response to NaCl stress: a meta-analysis. Front. Plant. Sci. 2014;5, 562. DOI: 10.3389/fpls.2014.00562.

[39] Shoresh M, Harman GE, Mastouri F. Induced systemic resistance and plant responses to fungal biocontrol agents. Annu Rev Phytopathol 2010;48:21-43. DOI: https://doi.org/10.1146/annurevphyto-073009-114450.

[40] Kiss L. A review of fungal antagonists of powdery mildews and their potential as biocontrol agents. Pest Manag Sci. 2003;59:475-483. DOI: https://doi.org/10.1002/ps.689.

[41] Harman GE, Howell CR, Viterbo A, Chet I, Lorito M. Trichoderma species — opportunistic, avirulent plant symbionts. Nat Rev Microbiol. 2004;2:43-56. DOI: https://doi.org/10.1038/nrmicro797.

[42] Fadiji AE, Babalola OO. Elucidating Mechanisms of Endophytes Used in Plant Protection and Other Bioactivities with Multifunctional Prospects. Front. Bioeng. Biotechnol. 2020;8:467. DOI: 10.3389/fbioe.2020.00467.

[43] Schachtman DP, Reid RJ, Ayling SM. Phosphorus uptake by plants: from soil to cell. Plant Physiol. 1998;116, 447-453. DOI: 10.1104/pp.116.2.447.

[44] Plassard C, Becquer A, Garcia K. Phosphorus Transport in Mycorrhiza: How Far Are We? Trends Plant Sci. 2019;24(9):794-801. DOI: 10.1016/j.tplants.2019.06.004.

[45] Smith SE, Smith FA. Roles of arbuscular mycorrhizas in plant nutrition and growth: new paradigms from cellular to ecosystem scales. Annu. Rev. Plant Biol. 2011;62, 227-250. DOI:10.1146/annurev- arplant-042110-103846.

[46] Bucher M. Functional biology of plant phosphate uptake at root and mycorrhiza interfaces. New Phytol. 2007;173,11-26. DOI:10.1111/j.1469-8137.2006.01935.x.

[47] Castellanos-Morales V, Villegas J, Wendelin S, Vierheiling H, Eder R, Cardenas-Navarro, R. Root colonization by the arbuscular mycorrhizal fungus Glomus intraradices alters the quality of strawberry fruit (Fragaria ananassa Duch.) at different nitrogen levels. J. Sci. Food Agric. 2010;90, 1774-1782. DOI: 10.1002/jsfa.3998.

[48] Hart M, Ehret DL, Krumbein A, Leung C, Murch S, Turi C, Franken P. Inoculation with arbuscular mycorrhizal fungi improves the nutritional value of tomatoes. Mycorrhiza. 2015;25, 359-376. DOI: 10.1007/s00572-014-0617-0.

[49] Baslam M, Garmendia I, Goicoechea N. Arbuscular mycorrhizal fungi (AMF) improved growth and nutritional quality of greenhouse grown lettuce. J. Agric. Food Chem. 2011;59, 5504–C5515. DOI: 10.1021/jf200501c.

[50] Rouphael Y, Franken P, Schneider C, Schwarz D, Giovannetti M, Agnolucci M. Arbuscular mycorrhizal fungi act as bio-stimulants in horticultural crops. Sci. Hort. 2015;196, 91-108. DOI: 10.1016/j.scienta.2015.09.002.

[51] Leifheit EF, Veresoglou SD, Lehmann A, Morris EK, Rillig MC. Multiple factors influence the role of arbuscular mycorrhizal fungi in soil aggregation-a meta-analysis. Plant Soil. 2014;374, 523-537. DOI: 10.1007/s11104-013-1899-2.

[52] Singh PK, Singh M, Tripathi BN. Glomalin: an arbuscular mycorrhizal fungal soil protein. Protoplasma. 2013;250, 663-669. DOI: 10.1007/s00709-012-0453-z.

[53] Wilson GWT, Rice CW, Rillig MC, Springer A, Hartnett DC. Soil aggregation and carbon sequestration are tightly correlated with the abundance of arbuscular mycorrhizal fungi: results from long term field experiments. Ecol. Lett. 2009;12, 452-461. DOI: 10.1111/j.1461-0248.2009.01303.x.

[54] Gutjahr C, Paszkowski U. Multiple control levels of root system remodeling in arbuscular mycorrhizal symbiosis.

Front. Plant Sci. 2013;4:204. DOI: 10.3389/fpls.2013.00204.

[55] Cameron KC, Di HJ, Moir JL. Nitrogen losses from the soil/plant system: a review. Ann. Appl. Biol. 2013;162, 145-173. DOI: 10.1111/aab.12014.

[56] Cavagnaro TR, Bender SF, Asghari HR, van der Heijden MGA. The role of arbuscular mycorrhizas in reducing soil nutrient loss. Trends Plant Sci. 2015;20, 283-290. DOI: 10.1016/j.tplants.2015.03.004.

[57] Clark RB, Zeto SK. Mineral acquisition by arbuscular mycorrhizal plants. J. Plant Nutr. 2000;23, 867-902. DOI: 10.1080/01904160009382068.

[58] Solaiman ZM, Mickan B. Mycorrhizal fungi: use in sustainable agriculture and land restoration. Springer, Berlin. 2014;41. DOI: https://doi.org/10.1007/978-3-662-45370-4_1.

[59] Emara HA, Nower A, Hmza E, Saad M, El Shaib F. Role of Mycorrhiza as Biofertilization of Banana Grand Naine on Nursery Stage. Int.J.Curr. Microbiol.App.Sci. 2018;7(10): 805-814. DOI: https://doi.org/10.20546/ijcmas.2018.710.089.

[60] Vosátka M, Albrechtová J, Patten R. The international market development for mycorrhizal technology. In: Varma A (ed) Mycorrhiza: state of the art, genetics and molecular biology, eco-function, biotechnology, eco-physiology, structure and systematics. Springer, Heidelberg 2008;419-438.

[61] Urgiles N, Loján P, Aguirre N, Blaschke H, Gunter S, Stimm B, Kottke I. Applicationofmycorrhizalrootsimproves growth of tropical tree seedlings in the nursery: a step towards reforestation with native species in the Andes of Ecuador. New For. 2009;38:229-239. DOI: https://doi.org/10.1007/s11056-009-9143-x.

[62] Zahawi RA, Eckert C, Chaves-Fallas JM, Schwanitz L, Rosales JA, Holl KD. The effect of restoration treatment soils and parent tree on tropical Forest tree seedling growth. Open For Sci J. 2015;5:154-161. DOI: https://doi.org/10.4236/ojf.2015.52015.

[63] Gohre V, Paszkowski U. Contribution of the arbuscular mycorrhizal symbiosis to heavy metal phytoremediation. Planta. 2006;223, 1115-1122. DOI: 10.1007/s00425-006-0225-0.

[64] Gonzalez-Chavez MC, Carrillo-Gonzalez R, Wright SF, Nichols KA. The role of glomalin, a protein produced by arbuscular mycorrhizal fungi, in sequestering potentially toxic elements. Environ. Pollut. 2004;130, 317-323. DOI: 10.1016/j.envpol.2004.01.004.

Chapter 4

Production of Vegetable Crops by Using Arbuscular Mycorrhizae

Ozlem Altuntas

Abstract

In modern agriculture, application of beneficial microorganisms has become more reliable and alternative source to reduce the application of pesticides. Several studies demonstrate that the beneficial microorganisms like arbuscular mycorrhizal (AM) fungi, Pseudomonas species, Trichoderma species etc. increase the plant growth and their and also improve the quality of soil. Additionally, these microorganisms increase the resistance of host plants against biotic and abiotic stresses. In the present chapter; vegetable crops in horticultural systems were focused. Most of the vegetable crop form symbiotic relationship with mycorrhiza acting as a bridge for the flow of energy and matter between plants and soils. The symbiotic relationship includes most species of vegetables and some species of fungi that have great relevance to soil ecosystem functions, especially nutrient dynamics, microbial processes, plant ecology, and agriculture. AMF can improve the nutrient and water uptake, induce tolerance of abiotic and biotic stress of their host plants. In the sustainable agriculture, the association of soil microorganisms with plant roots can also be exploited and in this way improve plant growth and productivity under normal and stressful environment. As a result, mycorrhizae improves plant growth, root structure development and crop yield and quality in almost any ambient condition. In addition, another benefit of mycorrhizae is that plants are resistant to diseases. it is concluded that arbuscular mycorrhizal infused pepper seedlings have high yield and quality. And also arbuscular mycorrhizae can be recommended for high yield and quality crop.

Keywords: Arbuscular mycorrhiza, vegetables, plant growth, nutrient uptake, yield

1. Introduction

Industrialization and rapid population growth, especially after World War II, caused significant environmental problems around the world. Among these problems, the most significant one was the hunger. In order to overcome this problem, different opinions have been put forward. One of them was to acquire new areas to agriculture while the other was to obtain maximum yield per area. Since the first suggestion was not easy to practicable, second one shined out as an important opportunity. As the world population grows, providing the necessary food for people to feed will increase the demand for agricultural production, which will be the biggest challenge facing agriculture. To meet this challenge, there is a need to focus on the soil biological system in the farmed land and the agricultural ecosystem as a whole. When we look at the current situation; although the food produced is not insufficient, there is a problem in distributing it to the regions in need. As a result of the problem arising from this distribution injustice (there are regions with hunger problems in the world), it has led to high input agriculture and green revolution for higher yields. In the green revolution, high inputs were used for high efficiency, injustice in the distribution of products to the world continued, while there was an excess of wasted food in some regions, the need for food continued in hungry regions. However, the constant and alarming increase of the human population still threatens the world's food security. Therefore, it is thinked that a second green revolution will be needed to increase food production by about 50% in the coming years [1, 2]. Moreover, the use of chemical fertilizers has theoretically reached its maximum use and there will be no yield increase due to the use of fertilizers [3, 4].

It is becoming increasingly clear that while increasing the yield by applying more chemical fertilizers to the soil, the soil and plants cannot maintain a healthy production for a long time. Because indiscriminate and over-application of chemical fertilizers poses a danger to human and environmental health, agronomists have sought alternative strategies that can ensure productivity while maintaining soil health. This new concept of agriculture, often referred to as "sustainable agriculture", requires agricultural practices that are environmentally friendly and maintain the long-term ecological balance of the soil ecosystem. In this context, the use of biofertilizers (beneficial microorganisms) in agriculture constitutes an environmentally friendly alternative to other applications of mineral fertilizers. Continuous investigation of the natural biodiversity of soil microorganisms and optimization of microbial interactions in the rhizosphere are prerequisites for the development of more efficient microbial inoculants. In agricultural production, in addition to providing sufficient food for the increasing human population, the quality of agricultural products, healthy, ecologically compatible, environmentally friendly techniques are increasing. Application of beneficial microorganisms, is an important techniques that improve the ecosystem, soil and human health.

For example, excessive use of nitrogen fertilizers causes nitrate accumulation, especially in green leaf-eaten crops, and contamination of groundwater by leaching of nitrogen fertilizers. The reduction or replacement of chemical fertilizers with the use of beneficial microorganisms has been proven by studies [5–7]. Since beneficial microorganisms fulfill important ecosystem functions for plants and soil, both healthy and high quality agricultural production and reduction of chemical input use can also play a key role in preventing yield reduction [8–13]. Moreover, in modern agriculture, many plant species traditionally produced due to the use of chemicals are susceptible to diseases. Stimulation of plant growth and crop protection can be improved by the direct application of a number of microorganisms known to act as bio-fertilizers and/or bio-preservatives. In addition, the production of metabolites related to root development and pathogen control (phytohormones, antimicrobials, antibiotics) and their direct effects on some metabolic activities, plant nutrients and water can be counted as their most obvious benefits. Although it has been repeatedly demonstrated over the last 150 years that bacteria and fungi promote plant growth and suppress plant pathogens, this knowledge has not been extensively used in agricultural biotechnology [14].

The second most common microorganism in the soil is fungus. It is the most preferred and studied group of soil fungi, which are mostly related to photosynthetic plants as mycorrhizal symbiotics. Mycorrhizae represent a vital component in plant ecosystems: They are widely distributed in natural and agricultural environments and are found in more than 80% of land plants, liverworts, ferns, woody gymnosperms and angiosperms and grasses. Providing an effective nutrient and water uptake, resulting in increased yield and resistance to environmental stresses (biotic and abiotic) most land plants need to be associated with mycorrhizal fungi. The use of plant - mycorrhiza symbiosis in natural and agronomic environments has high environmental and economic value. Mycorrhiza; It is a term derived from the Greek words mykes and rhiza, meaning mushroom and root respectively [15]. It was first used in 1885 by Albert Bernhard Frank. Mycorriza expresses a symbiotic life between soil fungi and plant roots [16]. Arbuscular mycorrhiza (AM) replaced the earlier term "vesicular–arbuscular mycorrhiza" (VAM) because not all endomycorrhizae of this type develop vesicles, but all form arbuscules.

Mycorrhizae are found in many environments and their ecological success is due to their wide variety. About 6000 species of mycorrhizal species in *Glomeromycotina*, *Ascomycotina* and *Basidiomycotina* have been recorded, and the use of molecular techniques increases this number. The taxonomic position of plant and fungal partners defines the mycorrhiza species; for this the main distinction is between endomycorrhizaes and ectomycorrhizaes. With the symbiosis ectomycorrhizae (ECMs) in the roots of trees and shrubs, hyphae remain extracellular and cause significant changes in root morphogenesis. In addition, ectomycorrises cause only subtle changes in epidermal or cortical cells [17]. In the endomycorrhizae, namely the arbuscular (AMs), ericoid, and orchid mycorrhizae, hyphae penetrate the stem cells to form an intracellular symbiosis independent of the plant host. While AMs are common among various plant taxa [18], the ericoid and orchid mycorrhizae are restricted to the family Ericales and Orchidaceae, respectively [19]. Arbuscular mycorrhizal (AM) fungi improve soil structure and aggregate stability [20]. Therefore, it can be expected to increase water absorption and plant nutrient uptake by plants in the treated soil, which may increase plant growth [20, 21]. Arbuscular mycorrhizal (AM) fungi, which are in symbiotic relationship with the roots of the majority of land plants, increase the nutrient-absorbing root surface area in the host plant through external myceliums [22]. In other words, root surface area increase is that mycorrhizae develop an extramatric mycelium, which in turn increases the plant nutrient absorption sites of the roots [23]. Since arbuscular mycorrhizal fungi are obligate endosymbions and live with carbohydrates derived from stem cells, all soil factors affecting plant growth and physiology will also alter fungal activity and thus affect the structure and functioning of bacterial communities [24]. It is now well understood that arbuscular mycorrhizal fungi alter root functions [25], alter the carbohydrate metabolism of the host plant, and affect rhizosphere populations [26]. Microorganisms in areas where hyphae of arbuscular mycorrhizal fungi extend may affect mycorrhizal functions such as nutrient and water uptake by arbuscular mycorrhizal fungi.

During intergenerational interactions, arbuscular mycorrhizal fungi improve the phosphate nutrition of plants by using the available phosphorus in the soil due to the large root surface area and high affinity phosphate uptake mechanisms created by the hyphae [27, 28]. The role of arbuscular mycorrhizal fungi in improving plant phosphate nutrition and their interactions with other soil biota has been investigated with reference to host plant growth, there is research on the organic acid production of arbuscular mycorrhizal fungi that can dissolve insoluble mineral phosphate [29, 30]. AMF has a number of well-documented effects on plant nutrition ([31], new literature), it is worth highlighting the potential role of AMF in micronutrient uptake in particular [32, 33] because of its important implications for the nutritional value of plant products. In addition, AMF potentially contributes to increased drought resistance of the crop by improving plant-water relationships through a variety of mechanisms [19]. In addition, AMF may interact with beneficial microorganisms such as phosphate-dissolving bacteria [34], with potential beneficial contributions to the nutrient cycle and plant nutrition. At the ecosystem scale, AMF gains importance with its effects on soil aggregation in soils where organic matter is the main binding agent. Soil aggregation has important implications for carbon storage [35, 36]. In addition to increasing water and nutrient intake in the soil, AMF, which provides carbon storage in the soil, is one of the main determinants of soil quality. AMF plays important roles in agroecosystems, including the participation of extra radical mycelium in promoting soil aggregation. Among these functions is their role in soil aggregation, hypothesized to be partly mediated by a proteinaceous compound released by an actively growing AMF mycelium in the soil: glomalin [37, 38]. This proteinaceous compound, which was operationally identified and extracted from soil as glomalin-associated soil protein [39], is highly associated with an important soil parameter and total water stability [40].

In order to ensure desired yields in vegetable cultivation, chemical fertilizers are applied in excess amount. In addition, more fertilizer are used in greenhouse vegetable growing because of the 2–3 times higher yield and the longer production season. There is a greater need for farmyard manure in greenhouse and field vegetable growing than other production systems. However, it is quite hard to supply large quantites of farmyard manure. For this reason, the need for plant nutrition instead of organic fertilizer in the soil is generally provided by chemical fertilizers. It is known that mineral fertilizer applications, especially nitrogen, are washed from the soil profile and cause pollution in ground waters. It is also believed that chemical compounds contribute to the greenhouse effect and the ozone layer under certain conditions. As a result of these negativities, agricultural practices that are friendly to the environment and which do not disrupt the ecological balance into the soil have been needed. In this context, biological fertilizers, plant stimulants and biological pesticides have been considered as resources that are able to meet the nutrients needed by the plant. Microorganisms to be used as biological fertilizers; should be simple to apply cheap, have high metabolic activity and be able to store for a long time.

As mentioned earlier, the use of excessive agricultural inputs to solve the hunger problem, which is the result of increased population, corrupts food and living quality. For this reason, organic farming has become an important part of the world and researchers have done a lot of study on this subject. However, the limited agricultural inputs that can be used in organic farming make plant nutrition difficult in this production system. In this case, the use of bio-fertilizers for their many positive effects on plants can be an alternative solution for this problem. In this review, the use of mycorrhizae (one of bio-fertilizers) for different purposes in vegetable growing was considered.

Vegetables are an important source for human nutrition. Turkey's geographical conditions enable the cultivation of all kinds of vegetables [16]. Greenhouse vegetable production in the Mediterranean countries are an important agricultural sector. Open field vegetable cultivation requires a long vegetation period, and high yield requires more intensive use of fertilizers in the greenhouse cultivation. Useful soil microorganisms are destroyed during the disinfection of greenhouse pests. These microorganisms do not exist in soilless cultivation media.

2. Use of mycorrhiza in soilless vegetables cultivation in the greenhouses

The effect of mychorrhizal inoculation with two species (*Glomus clarum* and *Glomus caledonium*) and three different inoculation treatments (sowing, transplanting and sowing + transplanting) were applied on pepper hydroponically grown on perlite medium. *G. clarum* and *G. caledonium* increased 29% and 21% respectively with respect to the control plants (**Table 1**). *G.clarum* was more effective on pepper yield. As seen in the **Figure 1**, plant growth and development especially root growth was excellent in plants inoculated with mycorrhizae. Mychorrhizae treatments increased pepper yield [41].

Dasgan et al. [42] studied soilless grown tomatoes inoculated with mycorrhizae in a plastic greenhouse (**Figure 2**). The substrate 1:1 perlite + cocopeat and nutrient

Experiment	Control	*G. caledomium*	*G. clarum*
Sowing (S)	839.92	912.38	1071.25
Transplanting (T)	839.92	964.67	989.71
S + T	**839.92**	**1076.58**	**1116.54**

Table 1.
The effect of mycorrhizae on the yield of peper plants at spring season (g plant^{-1}).

Figure 1.
Effect of mychorrhiza on the growth of pepper plants.

Figure 2.
Effect of mychorrhiza on the growth of tomato plants.

solution (full strength nutrients, the nutrient solution contained 20% and 40% and 60% reduced nutrients) were used. The yield was increased by mychorrhizae. The mycorrhizae along with nutrient solution responded differently. The higher yield was obtained in 60% nutrient solution (**Table 2**).

Treatments	Total yield
100% nutrient + M	9.65 c
80% nutrient + M	11.65 b
60% nutrient + M	**13.40 a**
40% nutrient + M	11.15 b
P	0.0035
LSD 0.005	1.39

Table 2.
The effect of nutrients and mycorrhizae on the yield of tomato plants (kg m^{-2}).

G. Fasciculatum was applied on tomato variety M19 and perlite was used as the substrate [43]. The mychorrizae use in soilless cultivation increases the tomato fruit yield. The highest yield (19.5 kg m^{-2}) was produced with the treatment under Open (M+) system (**Table 3**). The mychorrizal colonization in the open or closed systems affected the tomato yield. Higher fruit production was found for the mycorrhizal versus the non-mycorrhizal plants in both closed and open systems. Closed (M+) plants and Open (M+) plants produced 6.7% and 5.0% of higher yields, respectively, than those of the Closed (M-) and Open (M-) plants.

Yılmaz and Gül [44] studied the effect of mycorrhizae and phosphorus on the growth of eggplant (**Figure 3**). The cultivar Phaselis F1, and *Glomus caledonium* and the pumice were used. Among the 3 different phosphorus (15, 30 and 45 ppm) treatment, 15 ppm enhanced the yield along with mycorrhizal inoculation (**Table 4**).

Mycorrhizae fertilizer under the trade name 'Endo Roots Soluble' (ERS) was used in the experiment. The seeds of squash were directly sown into the substrate of perlite-cocopeat mixture in 1:1 ratio and cocktail mychorrhiza which contained *Glomus aggregatum, Glomus clarum, Glomus deserticola, Glomus etunicatum, Glomus intraradices, Glomus mosseae, Glomus monosporus, Glomus brasilianum* and *Gigaspora* [45]. The highest yield was obtained from the cocktail mychorrhiza +nutrients solution(80%) (**Table 5** and **Figure 4**).

Dere et al. [46] investigated the growth of cantaloupe melon at reduced mineral nutrients and mycorrhizal treatments ((1)100% full nutrition(control), (2) 100% full nutrition+mycorrhiza, (3) 80% nutrition,(4) 80% nutrition+mycorrhiza (5) 60% nutrition (6) 60% nutrition+mycorrhiza (7)40% nutrition, (8) 40% nutrition+mycorrhiza) (**Table 6**).

Mycorrhizal inoculation is an important for sustainable agriculture, as like chemical and biological factors in the soil strongly influence nutrient management.

Treatments	Yield
Closed (– M)	16.8 b
Closed (+ M)	18.0 b
Open (– M)	18.5 ab
Open (+ M)	**19.5 a**
P	0.047
LSD 0.005	1.758

Table 3.
The effect of mycorrhiza on the yield of tomato plant (kg m^{-2}) at colosed and open system.

Figure 3.
Effect of phosphorus and mycorrhiza on the soilless grown eggplant plants.

Treatments		1st year	2nd year
Mychorrhiza	—	4,67 b	4.88
	+		5.01
	LSD 0.05	0.22	ns
P doses	15 ppm	5.03	4.64
	30 ppm	5.09	5.25
	45 ppm	4.93	4.95
Mychorrhiza X P	- 15	4.51 d	4.33
	- 30	4.72 cd	5.42
	- 45	4.77 cd	4.89
	+ 15		4.95
	+ 30	5.46 ab	5.09
	+ 45	5.10 bc	5.00
	LSD 0.05	0.38	ns

Table 4.
The effects of phosphorus and mycorrhiza on the total yield of soilless grown eggplant plants (kg plant^{-1}).

Treatments	100% Nutrients	80% Nutrients	60% Nutrients
M +	383 b	**1019**	532
M -	797 a	984	598

Table 5.
The effect of mychorrhiza and reduced nutrients on the yield of squash plants under open soilless system (g m^{-2}).

Figure 4.
Role of cocktail mycorrhiza on the summer squash plants in the greenhouse.

Treatments	Total yield
100% N + M	9.64 b
100% N	8.28 bc
80% N + M	**12.44 a**
80% N	8.33 bc
60% N + M	6.09 cd
60% N	5.72 d
40% N + M	6.11 cd
40% N	5.93 cd

N: Nutrition, M: Mycorrhiza. Different letters on a column indicate significant differences according to Tukey's test ($P \leq 0.05$).

Table 6.
Effect of mycorrhizae on total yield of melon in the reduced nutrient levels.

For sustainable nutrient and water management, soil and crop management can be improved by using selected mycorrhizal spores [47] or by producing mycorrhizal inoculated on seedlings [48].

3. Conclusion

The cultivation of vegetables is a very important in the agricultural sector. For healthy vegetables production, the organic farming is one of the ways to bring stability and sustainability to agriculture. The complete elimination of chemical fertilizers is not possible. But the biofertilizers may reduce the chemical inputs. Mycorrhizae increase the plant growth and yield by providing water and nutrients. In conclusion, the mycorrhizae are important for the growth of agricultural crops as well as the health of ecosystem. Mycorrhizae inoculated plants can easily adapt to greenhouse and field conditions.

Author details

Ozlem Altuntas
Department of Horticulture, Faculty of Agriculture, Malatya Turgut Ozal University, Malatya, Turkey

*Address all correspondence to: ozlem.altuntas@ozal.edu.tr

References

[1] Leisinger, Klaus M. "Biotechnology and food security." CURRENT SCIENCE-BANGALORE- 76 (1999): 488-500.

[2] Vasil, Indra K. "Biotechnology and food security for the 21st century: a real-world perspective." Nature Biotechnology 16 (1998): 399-400.

[3] Abak, K., Dasgan, H. Y., Rehber, Y., & Ortaş, I. (2009, September). Effect of vesicular arbuscular mycorrhizas on plant growth of soilless grown musk-melon. In IV International Symposium on Cucurbits 871 (pp. 301-306).

[4] Ahmed, Benjamin. Economic Analysis Of Fertilizer Use In Maize Production In The Northern Guinea Savanna Of Nigeria. Diss. 1995.

[5] Burdman, S., E. Jurkevitch, and Y. Okon. "Recent advances in the use of plant growth promoting rhizobacteria (PGPR) in agriculture." Microbial interactions in agriculture and forestry (Volume II) (2000): 229-250.

[6] Dobbelaere, Sofie, Jos Vanderleyden, and Yaacov Okon. "Plant growth-promoting effects of diazotrophs in the rhizosphere." Critical reviews in plant sciences 22.2 (2003): 107-149.

[7] Frank, Birgit. Ueber die physiologische Bedeutung der Mycorhiza. 1888.

[8] Hermosa, Rosa, et al. "Plant-beneficial effects of Trichoderma and of its genes." Microbiology 158.1 (2012): 17-25

[9] Paul, E. A. and Clark, F. E. 1989. Soil biology and biochemistry. Academic Press, San Diego, CA.

[10] Raaijmakers, Jos M., et al. "The rhizosphere: a playground and battlefield for soilborne pathogens and beneficial microorganisms." Plant and soil 321.1-2 (2009): 341-361.

[11] Rhodes, L. H., and J. W. Gerdemann. "Phosphate uptake zones of mycorrhizal and non-mycorrhizal onions." New Phytologist 75.3 (1975): 555-561.

[12] Smith S., Read D. J. (2008) Mycorrhizal Symbiosis. Academic Press Publishers. London. p: 605.

[13] Whipps, John M. "Developments in the biological control of soil-borne plant pathogens." Advances in botanical research 26 (1997): 1-134.

[14] Berg, Gabriele. "Plant–microbe interactions promoting plant growth and health: perspectives for controlled use of microorganisms in agriculture." Applied microbiology and biotechnology 84.1 (2009): 11-18.

[15] Bardgett R. D. (2005) The Biology of Soil A Community and Ecosystem Approach. Oxford University Press, UK. p: 242.

[16] Muchovej R. M. (2009) Importance of Mycorrhizae for Agricultural Crops. The Institute of Food and Agricultural Sciences (IFAS). pp: 1-5.

[17] Auge R. M. (2001) Water relations, drought and vesicular-arbuscular mycorrhizal symbiosis. Mycorrhiza. 11: 3-42.

[18] Jeffries, P., Barea J. M. (2000) Arbuscular mycorrhiza – a key component of sustainable plant- soil ecosystems. In: Hock, B. (Ed.), The Mycota IX, Fungal Associations. Springer, Berlin. pp: 95-113

[19] Augé, R. M. 2001. Water relations, drought and vesicular-arbuscular mycorrhizal symbiosis. Mycorrhiza 11: 3-42.

[20] Leinhos V, Vacek O (1994) Biosynthesis of auxins by phosphate solubilizing rhizobacteria from wheat (Triticum aestivum) and rye (Secale cereale). Microbial Res 149:31-35

[21] Ortas, I. (2012). The effect of mycorrhizal fungal inoculation on plant yield, nutrient uptake and inoculation effectiveness under long-term field conditions. Field crops research, 125, 35-48.

[22] Olsson, Pål Axel. "Signature fatty acids provide tools for determination of the distribution and interactions of mycorrhizal fungi in soil." FEMS Microbiology Ecology 29.4 (1999): 303-310.

[23] Bolan, N. S. "A critical review on the role of mycorrhizal fungi in the uptake of phosphorus by plants." Plant and soil 134.2 (1991): 189-207.

[24] Azaizeh, H. A., Marschner, H., Römheld, V., & Wittenmayer, L. (1995). Effects of a vesicular-arbuscular mycorrhizal fungus and other soil microorganisms on growth, mineral nutrient acquisition and root exudation of soil-grown maize plants. Mycorrhiza, 5(5), 321-327.

[25] Marschner, Petra, David E. Crowley, and Richard M. Higashi. "Root exudation and physiological status of a root-colonizing fluorescent pseudomonad in mycorrhizal and non-mycorrhizal pepper (*Capsicum annuum* L.)." Plant and Soil 189.1 (1997): 11-20.

[26] Hobbie, Sarah E. "Effects of plant species on nutrient cycling." Trends in ecology & evolution 7.10 (1992): 336-339.

[27] Hayman, D. S. "The physiology of vesicular–arbuscular endomycorrhizal symbiosis." canadian Journal of Botany 61.3 (1983): 944-963.

[28] Humphreys, C. P., Franks, P. J., Rees, M., Bidartondo, M. I., Leake, J. R., & Beerling, D. J. (2010). Mutualistic mycorrhiza-like symbiosis in the most ancient group of land plants. Nature communications, 1(1), 1-7.

[29] Bagyaraj, D. J., Sharma, M. P., & Maiti, D. (2015). Phosphorus nutrition of crops through arbuscular mycorrhizal fungi. Current Science, 1288-1293.

[30] Lapeyrie, F. "Oxalate synthesis from soil bicarbonate by the mycorrhizal fungusPaxillus involutus." Plant and Soil 110.1 (1988): 3-8.

[31] Smith, S. E. and Read, D. J. 1997. Mycorrhizal symbiosis. Academic Press, San Diego, CA.

[32] Bati, C. B., Santilli, E., & Lombardo, L. (2015). Effect of arbuscular mycorrhizal fungi on growth and on micronutrient and macronutrient uptake and allocation in olive plantlets growing under high total Mn levels. Mycorrhiza, 25(2), 97-108.

[33] Marschner, H. 1995. Mineral nutrition of higher plants. 2nd ed. Academic Press, London, UK.

[34] Toro, M., Azcon, R. and Barea, J. M. 1998. The use of isotopic dilution techniques to evaluate the interactive effects of Rhizobium genotype, mycorrhizal fungi, phosphate-solubilizing rhizobacteria and rock phosphate on nitrogen and phosphorus acquisition by Medicago sativa. New Phytol. 138: 265-273.

[35] Jastrow, J. D. 1996. Soil aggregate formation and the accrual of particulate and mineral-associated organic matter. Soil Biol. Biochem. 28: 665-676.

[36] Six, J., Elliott, E. T. and Paustian, K. 2000. Soil macroaggregate turnover and microaggregate formation: a mechanism for C sequestration under no tillage agriculture. Soil Biol. Biochem. 32: 2099-2103.

[37] Rillig, M. C., Ramsey, P. W., Morris, S. and Paul, E. A. 2003. Glomalin, an

arbuscular-mycorrhizal fungal soil protein, responds to land-use change. Plant Soil 253: 293-299.

[38] Wright, S. F. and Anderson, R. L. 2000. Aggregate stability and glomalin in alternative crop rotations for the central great plains. Biol. Fertil. Soils 31: 249-253.

[39] Rillig, M. C. and Steinberg, P. D. 2002. Glomalin production by an arbuscular mycorrhizal fungus: a mechanism of habitat modification. Soil Biol. Biochem. 34: 1371-1374.

[40] Wright, S. F. and Upadhyaya, A. 1998. A survey of soils for aggregate stability and glomalin, a glycoprotein produced by hyphae of arbuscular mycorrhizal fungi. Plant Soil 198: 97-107.

[41] Ikiz, O.; Abak, K.; Daşgan, H.Y. and Ortaş, I. 2009. Effects of mycorrhizal inoculation in soilless culture on pepper plant growth. Acta Hort., 807: 533-593.

[42] Dasgan et al., 2013. "Using of Mychorriza in Soilless Grown Tomato for Saving Nutrients," 1st Central Asia Congress on Modern Agricultural Techniques and Plant Nutrition (AGRICASIA'2013)-Kyrgyzstan.

[43] Dasgan, H. Y., Kusvuran, S., & Ortas, I. (2008). Responses of soilless grown tomato plants to arbuscular mycorrhizal fungal (*Glomus fasciculatum*) colonization in re-cycling and open systems. African Journal of Biotechnology, 7(20).

[44] Yılmaz, E., & Gül, A. (2009). Effects of arbuscular mycorrhiza inoculation to soilless medium on eggplant (Solanum melongena L.) cultivation. Gaziosmanpasa University Journal of the Faculty of Agriculture, 2009, 26(2), 55-61.

[45] Dasgan, H. Y., Aydoner, G., & Akyol, M. (2010, August). Use of some microorganisms as bio-fertilizers in soilless grown squash for saving chemical nutrients. In XXVIII International Horticultural Congress on Science and Horticulture for People (IHC2010): International Symposium on 927 (pp. 155-162).

[46] Sultan, D. E. R. E., Coban, A., Akhoundnejad, Y., Ozsoy, S., & Dasgan, H. Y. (2019). Use of mycorrhiza to reduce mineral fertilizers in soilless melon (Cucumis melo L.) cultivation. Notulae Botanicae Horti Agrobotanici Cluj-Napoca, 47(4), 1331-1336.

[47] Ortas, I., 2003. Effect of selected mycorrhizal inoculation on phosphorus sustainability in sterile and Non-sterile Soils in the Harran plain in South Anatolia. J. Plant Nutr. 26, 1 17.

[48] Ortas, I., Sari, N., Akpinar, C., Yetisir, H., 2011. Screening mycorrhiza species for plant growth, P and Zn uptake in pepper seedling grown under greenhouse conditions. Sci. Hort. 128, 92 98.

Chapter 5

The Role of Mycorrhizae on Seedlings and Early Growth of Sugarcane

Wawan Sulistiono and Taryono

Abstract

The role of mycorrhizae in plant growth is well known, such as the ability to increase nutrient uptake, especially phosphate (P), drought tolerance, and resistance to pathogens. It is necessary to understand the application of arbuscular mycorrhizal technology in industrial plant production systems and their impact on agriculture systems. Large-scale nurseries of plantations require proper mycorrhizal application techniques. The relationship of mycorrhizal infection with plant yield (biomass) is known and in the next step, appropriate application time is needed to increase the effectiveness of mycorrhizae in plant growth and yield. Application of mycorrhizal inoculum was more effective in increasing the biomass of sugarcane stem weight to reach 61% with an increase in infection of 41.3%. In addition, the mycorrhizal application increases the root growth of sugarcane seedlings. The root growth promoting ability is important to increase the initial growth of plants after transplanting in dry land under the influence of drought stress, limited nutrients. The application of this technology is expected to increase plant growth, facilitate the maintenance and efficiency of cultivation on an industrial scale.

Keywords: mycorrhizae, seedlings, early growth, industrial crops, plantation

1. Introduction

Arbuscular mycorrhiza (*Zygomycetes*) is a symbiotic form of mutualism between fungal mycelium and higher plant roots. It is estimated that more than 80% of the higher terrestrial plants have a symbiosis with mycorrhizae [1]. The arbuscular mycorrhizae are known as vesicular-arbuscular mycorrhizae (VAM). VAM infects from outside the root into the root tissue and enter the root cells to form vesicles and arbuscules [2]. Arbuscules is a network of hyphae that penetrates between root cells and plasmalemma. Arbuscules help transport of nutrients to plant cells, especially P elements. In the root cells, the hyphae also form vesicles, small bubbles (granules) in the cytoplasm that contain lipids as a means of asexual reproduction of mycorrhizae. The vesicle plays a role during the reproduction, and when cells are ruptured at damage [2, 3]. Arbuscular mycorrhizae fungi (AMF) have very wide distribution in terrestrial ecosystems in terms of host plants, climate, and soil types [4]. Mycorrhizal infections will change the morphology of plant roots and nutrient absorption. This because their structures ensure the physical expansion of roots and absorption of nutrients from the soil and increase the flow of nutrients to plants [2, 3].

The role of mycorrhizae becomes important in sub-optimal land, dry land and for sustainable agriculture. Utilization of mycorrhizae, especially for plant growth, soil fertility and mitigation of drought stress by heat and climate change. Mycorrhiza becomes a component of future technology for sustainable agriculture [5, 6].

Mycorrhizae in agricultural land, especially sub-optimal land, functions to reduce soil erosion and leaching of nutrients. This condition is caused by the faster nutrient cycling mechanism. Besides, the absorption of nutrients is more due to the higher root surface area, which causes long-term soil fertility or soil productivity [5–7]. The presence of mycorrhizae in sub-optimal dry land of plantation crops is useful for renaturation and afforestation, namely stabilizing degraded land and eroding the soil surface [8]. In areas with high rainfall, plants in symbiosis with mycorrhizae also increase ecosystem repair by reducing the leaching of elements in the soil. Mycorrhizae will suppress the loss of nitrogen (N) and P elements by 40% and 50%, respectively in soil [9].

Mycorrhizal inoculation is important in dealing with drought stress and preparing plants for good growth in the field. Treat mycorrhizal inoculation on plantation seedlings to produce plants that have better root morphology and plant growth. These include root surface area in early coconut growth [10], root length, root diameter, root dry weight, and root dry weight ratio, root surface area, and shoot growth of sugarcane seedlings [11]. Likewise, mycorrhizae play a role in accelerating the growth or emergence of secondary roots in sugarcane seedling [12]. Mycorrhizae also appeared to have a significant role in increasing the growth of forest plant seedlings in the nursery, the increase in leaf chlorophyll content, photosynthesis rate, NPK content in root, stem and leaf compared to plants without mycorrhizal inoculation [13].

Thus mycorrhizal inoculation in plantation crops is needed as an effort to mitigate environmental stresses, both drought and high rainfall. In addition, the impact of mycorrhizal inoculum will increase the nutrients cycle in the soil, prevent excessive leaching of nutrients so that it plays a role in afforestation. In mycorrhizal inoculation, the inoculation time and dose are important. The optimal time of mycorrhizal inoculation is plantation seedlings in the nursery which will increase their colonization by 46% compared to field application [14].

The inoculation of mycorrhizae on seedlings of seedlings is expected to increase the morphological performance and plant physiological performance for early growth and morphological properties, increase the adaptive ability of plants to environmental stress. Based on the above, there are several important benefits for using mycorrhizal inoculation in the nursery of industrial plants.

2. Determinants of colonization and colonization patterns of plantation seeds

Before the inoculation of AMF on plantation seeds, it is necessary to know the determinants of colonization and the pattern of colonization. According to Sieverding [15], the process of colonization or infection progresses through 6 stages, namely: (1) pre-infection, at this stage, the spores are not yet active and AMF hyphae are in the soil; (2) penetration of the fungus to the roots. (3) arbuscules and vesicle formation. Arbuscule is formed after 2–5 days from penetration in the form of a strong band of hyphae growing around the cell plasmalemma. The vesicle at the tip of the hypha consists of lipids and fungal organs; (4) fungal elongation in roots and rhizosphere; (5) Spread of fungi to the soil. Hyphae grow out of the roots. The hyphae in the rhizosphere form the "external mycelium"; (6) culture of AMF structures into the form of resting spores on the external mycelium.

The elongation of fungi in the roots and rhizosphere consists of 3 stages, namely (a) slow phase, when infection to the target roots begins; (b) an exponential growth phase, maximum at 40 days after infection; (c) slowed growth phase, "plateau phase" balance [15]. Meanwhile, according to the observations of Sulistiono et al. [14] the colonization of sugarcane seedlings will experience a sharp increase at the age of 5–10 days after inoculation, then it will be constant at the age of 10–30 days after inoculation. An interesting point was also conveyed by Sulistiono et al. [14] that the tendency of AMF inoculation of sugarcane seeds in the nursery would result in higher colonization than inoculation carried out in the field when sugarcane at the age of 1–9 weeks after transplanting. However, after 9 weeks of age, the colonization rates of the two differences in inoculation time were similar. This indicates an equilibrium point for colonization or the development of infection at the root (**Figure 1**).

From the results of **Figure 1**, it shows that AMF inoculation in the nursery has several advantages, including:

1. Accelerate colonization when transplanting in the field

2. The AMF inoculated seeds has better growth of roots and shoots of plants

3. It has better adaptability to environmental stresses in the form of soil moisture and low nutrients, and diseases

4. Easy to apply

The higher colonization at the early growth of sugarcane was due to the effect of inoculation time which was applied in the nursery. AMF has infected and further developed which arbuscule and vesicle structures have formed [15] In the nursery, the colonization was optimal at 10–30 days after inoculation [12]. This is characterized by the formation of vesicules and arbuscules. The arbuscules and vesicles forms indicate symbiosis has occurred. This is because arbuscule are used for the transportation of nutrients from AMF to the root cells of host plant, especially P and vesicles are the reproductive organs of AMF and as a food reserve. One of the vesicles or arbuscules on the roots of sugarcane in the nursery as in **Figure 2**.

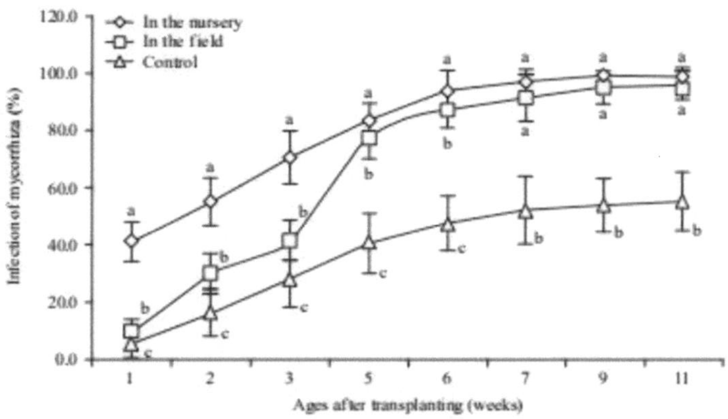

Figure 1.
The pattern of AMF colonization at different inoculation times.

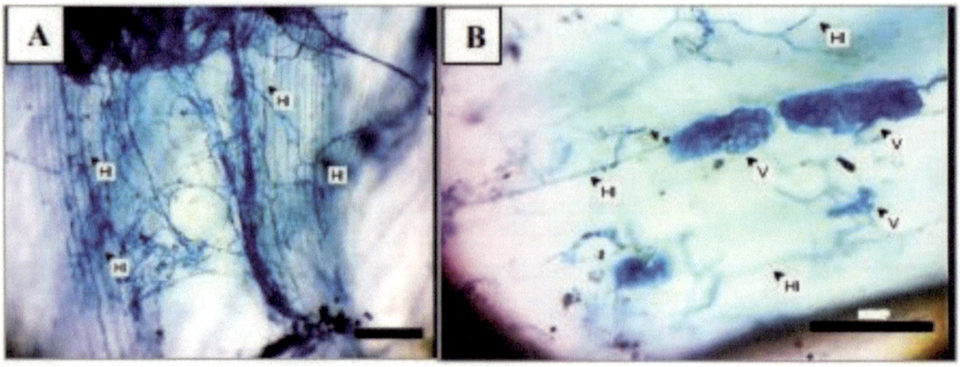

Figure 2.
Colonization on sugarcane bud chips at the age of 10 (A) and 20 days after inoculation (B). Hi: Hyphae; V: Vesicles. Scala bar: 10 µm. Objective 10×.

Therefore, AMF inoculation at seedling time results in an earlier infection growth process. This is indicated by the presence of hyphae structures since the age of 10 days after inoculation and vesicles at the age of 10 days after inoculation [12].

In the next stage, after the AMF structure is formed, it will accelerate the growth of secondary roots in sugarcane seeds, which was significantly different from the control (without inoculation) (**Table 1**).

Secondary roots in sugarcane seedlings are bigger roots and have a role to support the plant's upright and optimal absorption of nutrients. Thus, AMF inoculation in the nursery has the potential to increase plant growth (sugarcane) after transplanting. This is due to an increase in the number of secondary roots that are larger in diameter and also stronger [12].

Seeds/seedlings that have been inoculated with AMF in the nursery will have better growth properties in terms of leaf area, chlorophyll content, photosynthesis rate, and stem biomass in post-transplanting sugarcane seeds. This is because the application of AMF in the nursery increased the colonization by 41.3% at 7 days after transplanting and had the effect of increasing stem biomass from 11 to 61% (depending on sugarcane variety). This condition shows that there is a positive correlation between the rate of colonization and the weight of stem biomass, namely $r = 0.54$ [14].

AMF inoculation since seedling in forest plants (*Gleditsia sinensis* Lam) was also reported to increase seedling height, stem diameter, dry weight of seed biomass, chlorophyll content, photosynthetic rate, and NPK content in root, stem, and leaf tissue [13]. Likewise, inoculation of AMF in nurseries on tropical plant seeds was also reported to increase plant height, root diameter, and biomass [16] as well as N and P content of seedlings and root dry weight in forest plant seedlings [17].

AMF inoculation in perennial/industrial plant nurseries aims to prepare conditions for optimal growth factors, early symbiosis in the rhizosphere. This is because in 7–10 days the AMF structure has been formed, namely hyphae, vesicles, or arbuscule (**Figure 2**) [12]. With this difference in root symbiosis, plants can grow more optimally, uniformly, and faster. In this condition, it will provide an opportunity for healthy seed selection before transplanting in the field.

In the AMF inoculation treatment in the nursery, the things that need to be considered are the inoculum dose and the variety response. For plantation crops such as sugarcane, the optimal inoculation of AMF as much as 2 g of inoculum/seed or 7.8 spores/seedlings. This treatment resulted in significant root growth characteristics, shoot: root ratio and leaf P concentration compared to control [12].

Doses of AMF (g/seeds)	Number of secondary roots
0	1.60 b
1	2.80 ab
2	4.00 a
3	3.95 a

Remarks: Different letters in same column represent significant differences by Duncan's multiple range test at 5% level.

Table 1.
The effect of AMF doses level on number of secondary roots at the age of 40 days after inoculation and sowing.

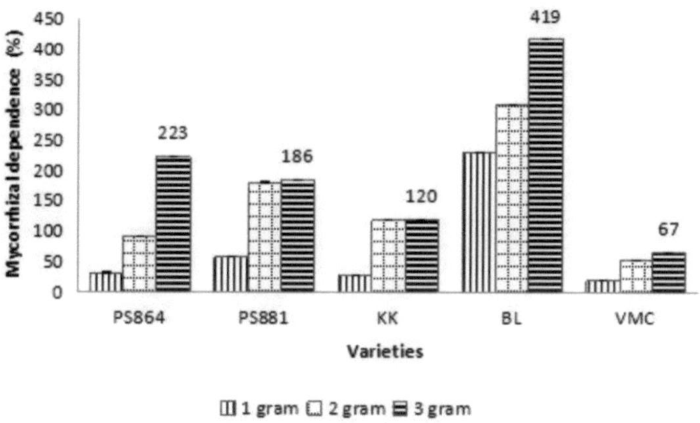

Figure 3.
Mycorrhizal dependence on several varieties of sugarcane.

The application of AMF inoculum is attempted in an optimal amount, in the right dose. The application of a higher dose will cause it to be less economical for a larger scale/volume.

AMF inoculation of industrial plant seedlings in the nursery needs to consider several limiting factors so that the colonization rate is optimal. Environmental factors are prepared since in the nursery. Environmental factors that determine the level of symbiosis with AMF, namely: (1) Light. Konvalinkova and Jansa [18] reported that the decreasing light intensity will decrease mycorrhizal growth (AMF) and decrease P transfer by AMF to host plants. This is because the availability of an energy source in the form of carbon is not sufficient for AMF and plant symbiosis. The light intensity which is only below 65% of a full-beam with 14–84 days shading time decreases the development of AMF in the root transfer of P elements from AMF to the host plant [18]. (2) soil temperature. The optimal soil temperature for AMF symbiosis with host plants is 20° C as indicated by the percentage of arbuscules and vesicles. An increase in temperature of 23–30° C causes a decrease in the arbuscules and vesicles formation [19]. (3) Elemental content of P, The addition of P into the soil showed a decrease in the percentage of mycorrhizal colonization [20]. (4) Host plants. The host plant is in the form of age, species, or variety [20, 21]. Different types of varieties respond to mycorrhizal inoculation as presented in **Figure 3**.

Figure 3 shows that genetically different varieties (sugarcane) have different mycorrhizal response [22]. These results can be used as the basis for selecting varieties for transplanting in the dry land. It can be concluded that:

1. Mycorrhizal inoculation to increase root and shoot growth
2. Preparation of a nursery that supports the symbiosis of AMF with plants
    a. Adjustment of nursery shade for colonization activities
    b. Setting the temperature of the media and nursery room for colonization
    c. Regulation of nutrient content, especially soil P, it should not excess.

3. Transplanting mycorrhizae inoculated seedlings for sustainable agriculture at adverse conditions

Mycorrhizal inoculation in plantation crops aims to promote good early growth and tolerate environmental stresses. A report shown that AMF inoculated seedlings

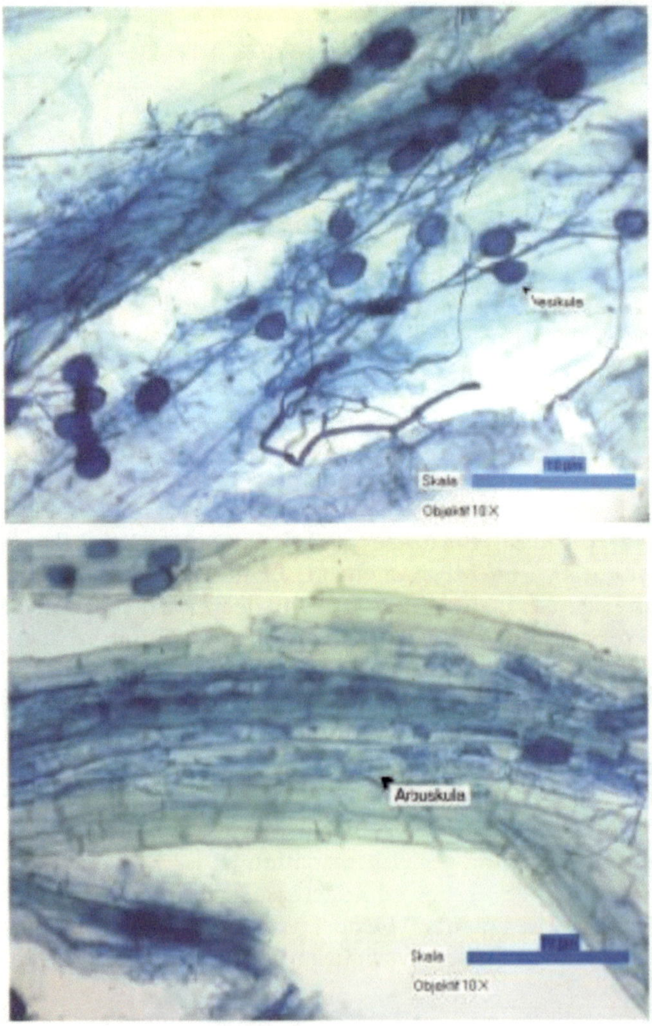

Figure 4.
The structure of arbuscules and vesicles and hyphae formed at the roots of plants infected with mycorrhizae. Field observations.

were then transplanted had increased leaf nitrate reductase activity (NRA) and root surface area in early coconut growth [10].

The increase in colonization with the formation of arbuscules and vesicles in early plant growth indicated that the symbiosis was optimal (**Figure 4**) [14]. This condition causes the host plant to obtain P elements from AMF transfer, more nutrient uptake by hyphae elongation and plant root structure, thus the plant experiences more optimal growth. Planting mycorrhizal inoculated plant seedlings is to increase nutrition in plants, especially P in cropping area [8], reduce nutrient loss due to leaching [6] so as to support sustainable agriculture [3, 8].

More stable nutrients available in mycorrhizae inoculated plant area resulted in increased soil productivity. Many nutrients are bound by the AMF structural system because there is a glomalin system. Glomalin as a glycoprotein forms chelates with inorganic P. Besides, the hyphae structure is more abundant which can directly absorb more nutrients, especially P [6]. Thus the AMF mycorrhizal inoculation treatment is a mitigation measure against climate change so that plants will continue to grow and survive.

4. Conclusion

AMF inoculation on seedlings is increased the root and shoot growth as well as increased the colonization. The nursery location is adapted for AMF growth and symbiosis. Plants infected with mycorrhizae had better growth (roots and shoots) ability after transplanting the seedlings. Besides that, it can reduce nutrient loss and maintain soil fertility so that it is an effort to mitigate climate change.

Acknowledgements

This work was supported by a grant from the Institute of research and community service, Lembaga Penelitian dan Pengabdian Kepada Masyarakat (LPPM) Universitas Gadjah Mada Yogyakarta Indonesia (UGM/396/LIT/2014). Authors thank for Prof. Dr. Prapto Yudono, Prof. Dr. Irham. Assesment Institute for Agricultural Technology of North Maluku (DIPA 2018)-Indonesian Agency for Agricultural Research and Development-Ministry of Agriculture. Authors thank for Dr. Andriko Noto Susanto, Dr. Bram Brahmantiyo and Dr. Abdul Wahab for the research facility.

Conflict of interest

I declare that I have no conflict of interest as an author on the financial and intellectual processes of the entire manuscript.

Author details

Wawan Sulistiono[1*] and Taryono[2]

1 Assessment Institute for Agriculture Technology (BPTP) of North Maluku, Indonesian Agency for Agricultural Research and Development, Ministry of Agricultural, Sofifi, Tidore Kepulauan City, Indonesia

2 The Center of Agrotechnology Innovation (PIAT UGM), Universitas Gadjah Mada, Kalitirto Berbah, Sleman, Yogyakarta, Indonesia

*Address all correspondence to: tionojanah@gmail.com

References

[1] Lee EH, Eo JK, Ka KH, Eom AH. Diversity of arbuscular mycorrhizal fungi and their roles in ecosystems. Mycobiology. 2013;**41**(3):121-125. DOI: 10.5941/MYCO.2013.41.3.121

[2] Quilambo OA. The vesicular-arbuscular mycorrhizal symbiosis. African Journal of Biotechnology. 2003;**2**(12):539-546

[3] Delian E, Chira A, Chira L, Arbuscular Mycorrhizae SE, Overview A. South west J Hortic. Biol Environ. 2011;**2**(2):167-192

[4] Begum N, Qin C, Ahanger MA, Raza S, Khan MI, Ashraf M, Ahmed N and Zhang L. Role of Arbuscular mycorrhizal fungi in plant growth regulation: Implications in abiotic stress tolerance. Frontiers in Plant Science. 2019;**10**:1068. DOI: 10.3389/fpls.2019.01068

[5] Solaiman ZM, Mickan B. Mycorrhizal fungi: Use in sustainable agriculture and land restoration. Soil Biology. 2014;**41**. DOI: 10.1007/978-3-662-45370-4_1

[6] Parihar M, Meena VS, Mishra PK, Rakshit A, Choudhary M, Yadav RP, Rana K, Bish JK. Arbuscular mycorrhiza: A viable strategy for soil nutrient loss reduction. Archives of Microbiology. 2019:1-14. DOI: org/10.1007/s00203-019-01653-9

[7] Casazza G, Lumini E, Ercole E, Dovana F, Guerrina M, Arnulfo A, Minuto L, Fusconi A, Mucciarelli M. The abundance and diversity of arbuscular mycorrhizal fungi are linked to the soil chemistry of screes and to slope in the Alpic paleo-endemic *Berardia subacaulis*. PLoS One. 2017;**12**(2):e0171866. DOI: 10.1371/journal. pone.0171866

[8] Chen M, Arato M, Borghi L, Nouri E, Reinhardt D. Benefificial services of arbuscular mycorrhizal fungi-from ecology to application. Front. Plant Science. 1270:9. DOI: 10.3389/fpls.2018.01270

[9] Inez-Garcıa LBM, Deyn GB, Pugnaire FI, Kothamasi D, van der Heijden MGA. Symbiotic soil fungi enhance ecosystem resilience to climate change. Glob Change Biol. 2017;**23**:5228-5236. DOI: 10.1111/gcb.13785

[10] Sulistiono W, Brahmantiyo B, Hartanto S, Aji HB, Bina HK. Effect of arbuscular mycorrhizal fungi and npk fertilizer on roots growth and nitrate reductase activity of coconut. Journal of Agronomy. 2020;**19**(1):46-53. DOI: 10.3923/ja.2020.46.53

[11] Sulistiono W, Taryono, Yudono P, Irham. Sugarcane roots dynamics inoculated with arbuscular mycorrhizal fungi on dry land. Journal of Agronomy 2017. 16: 101-114.. DOI: 10.3923/ja.2017.101.114

[12] Sulistiono W. Taryono, Yudono P, Irham. Application of arbuscular mycorrhizal fungi accelerates the growth of shoot roots of sugarcane seedlings in the nursery. Australian Journal of Crop Science. 2018;**12**(07):1082-1089. DOI: 10.21475/ajcs.18.12.07.PNE1001

[13] Wang J, Zhong H, Zhu L, Yuan Y, Xu L, Wang GG, Zai L, Yang L, Zhang J. Arbuscular mycorrhizal fungi effectively enhances the growth of *Gleditsia sinensis* lam. Seedlings under greenhouse conditions. Forests. 2019;**10**:567. DOI: 10.3390/f10070567

[14] Sulistiono W , Taryono, Yudono P, Irham. Early-Arbuscular Mycorrhizal Fungi-Application Improved Physiological Performances of Sugarcane Seedling and Further Growth. Journal of Agricultural Science.

2017;**9**(4):95-108. DOI: 10.5539/jas.v9n4p95

[15] Sieverding E. Vesicular-arbuscular mycorrhiza management in tropical agrosystem. Deutsche Gesellschaft für Technische Zusammenarbeit (GTZ) GmbH. In: Technical Cooperation-Federal Republic of Germany. 1991

[16] Urgiles N, Lojan P, Aguirre N, Blaschke H, Gunter S, Stimm B, Kottke I. Application of mycorrhizal roots improves growth of tropical tree seedlings in the nursery: A step towards reforestation with native species in the Andes of Ecuador. New Forests. 2009. DOI: 10.1007/s11056-009-9143-x

[17] Wulandari D, Saridi CW. Tawaraya K. Arbuscular mycorrhizal colonization enhanced early growth of *Mallotus paniculatus* and *Albizia saman* under nursery conditions in East Kalimantan, Indonesia. International Journal of Forestry Research. 2014;**898494**:1-8. DOI: org/10.1155/2014/898494

[18] Konvalinková T, Jansa J. Lights off for arbuscular mycorrhiza: On its symbiotic functioning under light deprivation. Frontiers in Plant Science. 2016; Front. Plant Sci. 7:782. doi: 10.3389/fpls.2016.00782

[19] Heinemeyer A, Fitter AH. Impact of temperature on the arbuscular mycorrhizal (AM) symbiosis: Growth responses of the host plant and its AM fungal partner. Journal of Experimental Botany. 2004;**55**(396):525-534. DOI: 10.1093/jxb/erh049

[20] Carrenho R, Trufem SPB, Bononi VLR, Silva ES. The effect of different soil properties on arbuscular mycorrhizal colonization of peanuts, sorghum and maize. Acta Botânica Brasílica. 2007;**21**(3):723-730

[21] Hindumathi A, Reddy BN. Occurrence and distribution of arbuscular mycorrhizal fungi and microbial flora in the rhizosphere soils of mungbean [*Vigna radiata* (L.) wilezek] and soybean [*Glycine maz* (L.) Merr.] from Adilabad, Nizamabad and Karimnagar districts of Andhra Pradesh state, India. Advance in Bioscience and Biotechnology. 2011;**2**:275-286. DOI: 10.4236/abb.2011.24040

[22] Sulistiono W. Pengembangan teknologi sistem pindah tanam bibit pada budidaya tebu (*Saccharum officinarum* L) lahan kering [Disertation]. Yogyakarta: Universitas Gadjah Mada; 2017

Chapter 6

Assessment of Biocontrol Potential of Arbuscular Mycorrhizal (*Glomus* spp.) against Damping-off Disease (*Rhizoctonia solani*) on Cucumber

Baker Diwan Getheeth Aljawasim, Hussein M. Khaeim and Mustafa A. Manshood

Abstract

Rhizoctonia solani is one of the most important causative agents of damping-off diseases on cucumber plants and significantly reduces their yield. *R. solani* possesses some characteristics, such as wide host range and unlimited survival in soil, which made it most difficult to control. Therefore, the research for a biocontrol agent will be valuable to control this disease. Two species of mycorrhizal fungi (*Glomus mosseae* and *Glomus clarum*) that were evaluated against the agent *R. solani* reduced the damping-off disease on the cucumber plant. Mycorrhizal-inoculated plants with both species showed a significant reduction in disease severity (*DS*), which were 21 and 25%, respectively, whereas the disease severity was 65% for non-inoculated plants. Furthermore, the effects of mycorrhizal fungi were evaluated against the growth parameters of cucumber plants. Plants inoculated with both species of mycorrhizal fungi showed a significant increase in both shoot dry weight and root dry weight compared with uninoculated plants. In conclusion, both mycorrhiza species could be an important tool to control soil-borne pathogens, increase plant's nutrients' absorption, and increase resistance to abiotic stresses.

Keywords: biological control, *Rhizoctonia solani*, arbuscular mycorrhiza, cucumber, damping-off diseases

1. Introduction

Rhizoctonia solani Kühn, the causative agent of damping-off disease in a variety of crop plants such as cucumber, is an economical important soil-borne pathogen [1, 2]. *R. solani* fungus is considered as a difficult pathogen to control due to several characters such as the great variability in the pathogen population, a wide host range, and long-term survival in soil [3]. Further, some cultural practices including the crop rotation, sanitation, and soil solarization with *R. solani* are not sufficiently effective because the pathogen is able to survive for many years in soil. The application of chemical pesticides, mainly methyl bromide, is the most reliable method to control *R. solani*; however, it causes serious risks including polluting the air, damaging the environment, building fungicides' resistance of pathogen, and harming the human health [4, 5]. Therefore, the biological control method becomes an important component of the disease management to increase crop production and food safety [6].

The biological control becomes an important target of many researchers in the field of biological and agricultural sciences [5]. Biocontrol agents use different mechanisms of action against fungal pathogens, such as antimicrobial compound production activity, mycoparasitism or hyperparasitism, cell wall-lytic enzyme activity, and the application of systemic resistance (ISR) activity [7]. In addition, some biocontrol agents are capable of improving some aspects of plant growth, such as the germination rate, shoot and root weight, nutrients' uptake, and yield [8].

Arbuscular mycorrhizal (AM) fungi have been known to form a symbiotic relationship with around 80% of vascular plants. The symbiotic relationship can provide the plant with many benefits, including enhancement of plant growth and germination rates, increasing supplement of water and nutrients [9, 10]. In return, the AM fungi are completely dependable on the nutrients that are coming from the living root system [9]. In addition, AM fungi have been known to increase the host's resistance to a wide range of fungal and bacteria pathogens, especially rot pathogens [11]. The aim of this study was to examine the influence of different species of arbuscular mycorrhizal (AM) fungi (*Glomus* spp.) to promote systemic resistance against the disease agent of damping-off disease (*R. solani* Kühn) on cucumber (*Cucumis sativus* L.).

2. Materials and methods

Infected samples were brought from cucumber plants with wilting, yellowing, and dwarfing symptoms from a field related to the College of agriculture, University of Al-Qadisiyah. The plants were washed with sterilized water to remove soil residues and were cut to small pieces. Then, the samples were sterilized with sodium hypochlorite (NaClO) 1% for 2 min, washed with sterilized water twice, and dried with filter papers. Nine petri dishes of potato dextrose agar (PDA) were inoculated with five pieces of the infected plants and incubated for 3 days at 25°C. Soil samples were diluted for pathogen isolation and the petri dishes were incubated at 27°C. Both plant and soil samples were kept in a refrigerator at 4°C and diagnosed using classification keys [12].

Isolated pathogens were stored at 4°C prior to analysis and incubated at 25°C for 3 days. From the colony edge, four populated agar disks (7 mm) were cut and mixed in a 250 ml flask containing 100 ml of potato dextrose broth and 25 mg of chloramphenicol [13]. Sterilized soils were separated on each pot (3 kg) and inoculated with 1 ml from pathogen broth culture, and sterilized water was used for the control. Then, all pots were irrigated and covered for 3 days. Cucumber seeds were disinfected with sodium hypochlorite (NaClO) 1% for 4 min and were planted in each pot. Germinated, not germinated seeds, and collapsed plants were recorded after 7 and 10 days for planting, and disease intensity was calculated as recommended [14]: 0 = no symptoms; 1 = seed rot, not germinated; 2 = brown rot on the stem base, plant is still standing; 3 = plant is wilted, laying on the ground; and 4 = plant is dead. *DS* was calculated from disease grades 0–3 using the following formula [15]:

$$DS = \frac{\Sigma(f * v)}{N * X} \times 100 \quad (1)$$

where *DS* = disease severity, *f* = infection class frequencies, *v* = number of plants of each class, *N* = total of observed plants, and *X* = highest value of the evaluation scale.

Cucumber seeds were surface-sterilized using 0.2% NaClO for 2 min and rinsed several times with distilled water. Arbuscular mycorrhizal (AM) fungi were obtained from the Iraqi Ministry of Sciences and Technology's laboratory. This mixture consists of propagated units of *Glomus clarum* (Nicol. Schenck) and *Glomus mosseae* (Nicol. Gerd) in a suspension form (1×10^6 unit L^{-1} concentration). *Glomus* spp. were identified and separated in two tubes by the experts at Iraqi Ministry of Sciences and Technology's laboratory. Six healthy seeds of cucumber were planted in each pot (25 cm in diameter), which contained 3 kg of sterilized soil (clay:sand, 2:1, v/v) into each pot. For mycorrhizal inoculum, each pot was inoculated with dilution of 5 ml of either *Glomus clarum* or *G. mosseae*/L^{-1} water twice at the beginning of cultivation and after 14 days. As controls, the pots were provided with no AM + no pathogen, AM only, and pathogen only. For the pathogen inoculum, 5 ml of spore suspension (*R. solani*) was added at the beginning of cultivation. Six treatments were conducted as the following: *Glomus clarum*, *G. mosseae*, *G. clarum* + *R. solani*, *G. mosseae* + *R. solani*, control, and control + *R. solani*. Four replicates were made for each treatment. In this study, all plants did not receive any fertilizer and were watered when necessary at outdoor conditions. The disease severity for each treatment was monitored and estimated as mentioned above [16].

When the plants emerged above the soil surface, five plants were harvested from each treatment after 5, 10, 15, and 20 days. The plants were washed with tap water to clean off soil particles. Fresh and dry weights were evaluated and recorded after drying the samples by a hot air oven at 60°C for 48 h until gaining constant weight [17].

3. Results and discussion

Five pathogens were isolated form the infected plants and soil. The fungal identification was performed according to the morphological characteristic as previously reported in literatures [18, 19]. Among five isolated pathogens, *R. solani* showed the highest disease severity (*DS*) on cucumber plants, which was about 63%, while *Penicillium* spp. showed the lowest disease severity (*DS*), which was about 8% (**Figure 1**). Therefore, *R. solani* was the most aggressive pathogen due to the suitable environment condition, and the availability of susceptible hosts and was used for all subsequent studies.

The effect of AM fungi against *R. solani* on cucumber plants was studied by the inoculation of cucumber plants with the AM, *G. mosseae* + *G. clarum*, which showed a significant reduction in the disease severity of damping-off compared with control (**Figure 2**). Disease severity (*DS*) of mycorrhizal plants was reduced by 46% and 41%, respectively. Furthermore, inoculated plants with mycorrhiza showed fewer symptoms compared with non-mycorrhizal plants. Disease severity in AM-inoculated plants with *G. mosseae* was about 20%, which was slightly less than AM-inoculated plants with *G. clarum* (**Figure 2**).

The effect of AM fungi on the growth parameters of cucumber plants was assessed by shoot dry weight and root dry weight. AM fungi-colonized plants had significantly increased shoot and root dry weights when compared with the non-mycorrhizal plants (**Table 1**). Cucumber plants, colonized with AM (*G. mosseae*), showed a slight increase in all growth parameters compared with the plant colonized with AM (*G. clarum*), which matches with our results on the *DS* experiment (**Table 1**).

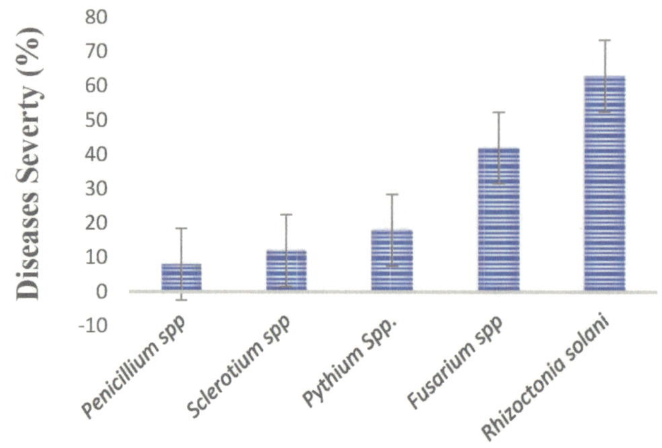

Figure 1.
Pathogenicity test for isolated pathogens against damping-off diseases on cucumber. Each column represents the mean of five replicates. Bars on the pillars represent standard error and LSD = 5.73 (P = 0.01).

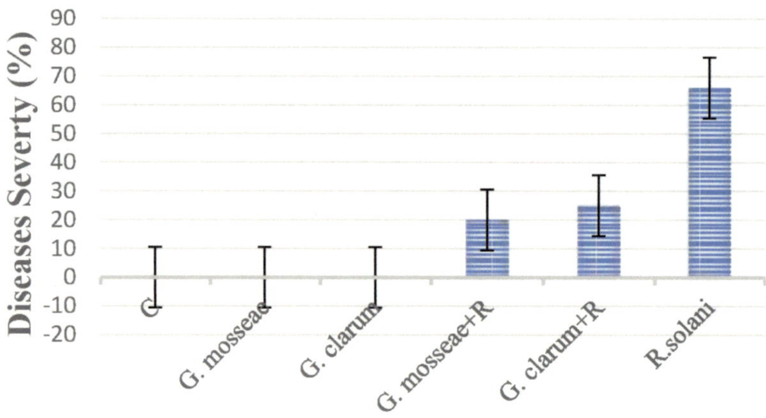

Figure 2.
Evaluation of arbuscular mycorrhizal (AM) fungi on the disease severity of damping-off diseases on cucumber. Each column represents the mean of four replicates. Bars on the pillars represent standard error and LSD (P = 0.01).

Mycorrhizal fungi are considered as ideal biocontrol agents due to some characteristics such as the ability to form a mutualistic symbiosis relationship with the roots of most vascular plant species [20]. Moreover, the plant-mycorrhiza relationship benefits the plant not only to control soil-borne pathogens but also to enhance the plant's resistance to various abiotic stresses and increases the nutrients' absorption [21].

In the present study, inoculated plant with mycorrhizal fungi reduces significantly the disease severity of *R. solani* pathogen, which may be attributed to increase the nutrients' status, reduce the direct competition for root space and resources with the pathogen, induce the plant's immunity to involve certain systemic mechanisms such as the systemic acquired resistance (SAR) and cell wall defenses, and enhance the production of defense compounds such as phenolics, -1,3-glucanase, and chitinolytic enzymes [9]. Additionally, inoculated plants with mycorrhizal fungi (*G. mosseae*) showed a lower disease severity than *G. clarum*, which may lead to a potential active control tool. Furthermore, the inoculation with mycorrhizal fungi increases both the root dry weight and shoot dry weight, which supports our hypothesis.

Mycorrhizal fungi play a main part in plant defense against pathogens and form a mutual relationship with plants. In summary, both mycorrhiza species could be

Treatment	Shoot dry weight (g/plant)				Root dry weight (g/plant)			
	5 days	10 days	15 days	20 days	5 days	10 days	15 days	20 days
Control	0.5	0.8	0.9	1.1	0.2	0.4	0.7	0.9
Control + *R. solani*	0.1	0.3	0.4	0.5	0.08	0.1	0.2	0.3
Glomus clarum	0.4	0.6	0.7	1.2	0.15	0.3	0.6	0.8
G. mosseae	0.6	0.7	0.8	1.1	0.2	0.4	0.7	0.9
G. clarum + *R. solani*	0.3	0.5	0.6	0.9	0.15	0.2	0.5	0.6
G. mosseae + *R. solani*	0.4	0.6	0.8	1	0.2	0.3	0.6	0.7

Table 1.
Evaluation of AM fungi on the growth parameters of cucumber plants.

an important tool to control soil-borne pathogens, increase plant nutrient absorption, and increase resistance to abiotic stresses. In future research, specific systemic mechanisms of mycorrhiza fungi against pathogens should be investigated more.

Acknowledgements

The research was supported by University of Muthanna, Iraq. The authors acknowledge the Ministry of Sciences and Technology in Iraq for providing them with the isolates of arbuscular mycorrhizal (AM) fungi to complete their research.

Author details

Baker Diwan Getheeth Aljawasim[1*], Hussein M. Khaeim[2] and Mustafa A. Manshood[1]

1 Department of Plant Protection, College of Agriculture, Al-Muthanna University, Iraq

2 Department of Soil Science and Water, College of Agriculture, University of Al-Qadisiyah, Iraq

*Address all correspondence to: baker.aljawasim@mu.edu.iq

References

[1] Saberi M, Sarpeleh A, Askary H, Rafiei F. The effectiveness of wood vinegar in controlling *Rhizoctonia solani* and *Sclerotinia sclerotiorum* in green house-cucumber. International Journal of Agricultural Sciences and Natural Resources. 2013;**1**(4):38-43

[2] Bartz FE, Cubeta MA, Toda T, Naito S, Ivors KL. An in planta method for assessing the role of basidiospores in *Rhizoctonia* foliar disease of tomato. Plant Disease. 2010;**94**(5):515-520

[3] Thakur M, Sahu NR, Tiwari P, Kotasthane A. Combination of azoxystrobin + difenocanazole provides effective management of sheath blight of rice caused by *Rhizoctonia solani*. IJCS. 2018;**6**(4):1682-1685

[4] Vinale F, Sivasithamparam K, Ghisalberti EL, Marra R, Woo SL, Lorito M. *Trichoderma*–plant–pathogen interactions. Soil Biology and Biochemistry. 2008;**40**(1):1-10

[5] Manganiello G, Sacco A, Ercolano MR, Vinale F, Lanzuise S, Pascale A, et al. Modulation of tomato response to *Rhizoctonia solani* by *Trichoderma harzianum* and its secondary metabolite harzianic acid. Frontiers in Microbiology. 2018;**9**:1966

[6] Justyna N, Magdalena S, Urszula M. *Trichoderma atroviride* enhances phenolic synthesis and cucumber protection against *Rhizoctonia solani*. Plant Protection Science. 2017;**54**(1):17-23

[7] Vinale F, Marra R, Scala F, Ghisalberti E, Lorito M, Sivasithamparam K. Major secondary metabolites produced by two commercial *Trichoderma* strains active against different phytopathogens. Letters in Applied Microbiology. 2006;**43**(2):143-148

[8] Liu K, McInroy JA, Hu C-H, Kloepper JW. Mixtures of plant-growth-promoting Rhizobacteria enhance biological control of multiple plant diseases and plant-growth promotion in the presence of pathogens. Plant Disease. 2018;**102**(1):67-72

[9] Jacott CN, Murray JD, Ridout CJ. Trade-offs in arbuscular mycorrhizal symbiosis: Disease resistance, growth responses and perspectives for crop breeding. Agronomy. 2017;**7**(4):75

[10] Smith SE, Smith FA, Jakobsen I. Mycorrhizal fungi can dominate phosphate supply to plants irrespective of growth responses. Plant Physiology. 2003;**133**(1):16-20

[11] Hoeksema JD, Chaudhary VB, Gehring CA, Johnson NC, Karst J, Koide RT, et al. A meta-analysis of context-dependency in plant response to inoculation with mycorrhizal fungi. Ecology Letters. 2010;**13**(3):394-407

[12] Domsch KH, Gams W, Anderson T-H. Compendium of Soil Fungi. Vol. 1. London: Academic Press Ltd.; 1980

[13] Jaiswal AK, Elad Y, Graber ER, Frenkel O. *Rhizoctonia solani* suppression and plant growth promotion in cucumber as affected by biochar pyrolysis temperature, feedstock and concentration. Soil Biology and Biochemistry. 2014;**69**:110-118

[14] Khan MR, Fischer S, Egan D, Doohan FM. Biological control of *Fusarium* seedling blight disease of wheat and barley. Phytopathology. 2006;**96**(4):386-394

[15] Abawi G, Widmer T. Impact of soil health management practices on soilborne pathogens, nematodes and root diseases of vegetable crops. Applied Soil Ecology. 2000;**15**(1):37-47

[16] Al-Askar A, Rashad Y. Arbuscular mycorrhizal fungi: A biocontrol agent against common. Plant Pathology Journal. 2010;**9**(1):31-38

[17] Manila R, Nelson R. Nutrient uptake and promotion of growth by arbuscular mycorrhizal fungi in tomato and their role in bio-protection against the tomato wilt pathogen. Journal of Microbiology and Biotechnology Research. 2017;**3**(4):42-46

[18] Sharma M, Gupta S, Sharma T. Characterization of variability in *Rhizoctonia solani* by using morphological and molecular markers. Journal of Phytopathology. 2005;**153**(7-8):449-456

[19] Guleria S, Aggarwal R, Thind T, Sharma T. Morphological and pathological variability in rice isolates of *Rhizoctonia solani* and molecular analysis of their genetic variability. Journal of Phytopathology. 2007;**155**(11-12):654-661

[20] Song Y, Chen D, Lu K, Sun Z, Zeng R. Enhanced tomato disease resistance primed by arbuscular mycorrhizal fungus. Frontiers in Plant Science. 2015;**6**:786

[21] Smith SE, Read DJ. Mycorrhizal Symbiosis. Academic Press; 2010

Chapter 7

Native Arbuscular Mycorrhizal Fungi and Agro-Industries in Arid Lands: Productions, Applications Strategies and Challenges

Bencherif Karima and Therrafi Samia

Abstract

Bio-fertilizers based on mycorrhizal fungi represent a natural way to enrich the soil in respect of environmental balance. Arbuscular mycorrhizal fungi (AMF) are the most common symbiotic association between terrestrial plants and microorganisms, which are known to improve plants development and growth, especially under stress conditions. The potential for application of AMF in agricultures is an agro-ecological approach to allow better use of soil nutrient reserves. That receives increasing consideration for their prospective application for sustainable agriculture. The present chapter aims to highlight the agro-industrial strategy of AMF bio-fertilizers production explaining agronomics, ecological and economic approaches and benefits. This study aims to focus on the importance of production of bio-fertilizers based on indigenous AMF strains and their role in improving soils enrichment, which will subsequently lead to improved production and agricultural yields on degraded arid soils.

Keywords: degraded areas, native inocula, industrial production strategies, agro-economic benefits, conventional method

1. Introduction

Soils are considered as a dynamic system that contains varieties of microorganisms such as bacteria, actinomycetes, and fungi [1]. According to this richness in microorganisms, the eco-biological value of soil is considered. Whereas, maintaining this favorable soil microflora is very important for soil sustainability [2].

In the other hand, arid lands constitute about 35% of world land areas and are characterized by rainfall insufficiency, higher temperatures and evapotranspiration, lower humidity, and a general rareness of vegetation cover [3]. In return, a large mass of world's population lives in these areas, which it is imperative to nourish them. They practice livestock grazing and irrigated agriculture that they try to modernize in order to obtain the best yield. However, the agricultural techniques used in recent decades (use of large quantities of chemical inputs, soil compaction, etc.) which have caused in addition to soil degradation the decrease or even elimination of certain beneficial microorganisms from most cultivated soils, which has contributed to the loss of productivity of these soils [4]. This destitution requires regular additions in order to revitalize the soil and restore its productivity [2]. But what type of fertility for the soil? In agronomy, the notion of fertilization includes application of various chemicals products such as NPK chemical formulation, pesticides and herbicides, which further degrade the soil and reduce their duration [5]. In fact, modern agriculture are based on heavy usage of chemical fertilizers and harmful pesticides on the crops, with destruction of sustainability of the agricultural systems, cost of cultivation soared at a high rate, income of farmers stagnated and food security and safety became an intimidating challenge with considerable reduction in soil health [6]. In the best of cases where there is an interest in ecological stability, animal and plant waste is applied to fertilize the soils [5]. Recently bio-fertilizers notion begins to emerge. Bio-fertilizers are biological fertilizers based on symbiotic microorganisms. They are mainly divided into two groups: bio-fertilizers based on symbiotic bacteria and bio-fertilizers based on mycorrhizal fungi [7].

Bio-fertilizers are biological fertilizers based on plant symbiotic microorganisms, they are defined as a substance composed of living microorganisms which when applied to seed, plant surfaces, or soil colonizes the rhizosphere or the interior of the plant and promotes growth by increasing the availability of primary nutrients to the host plant [6]. Usually, the bio-fertilizers are mainly divided into two groups: bio-fertilizers based on symbiotic bacteria and bio-fertilizers based on mycorrhizal fungi. Moreover, the presence in nature of bacteria solubilizing mineral elements is exploited; they are cultivated and used as a bio-fertilizer [4]. Furthermore, mycorrhizal fungi bio-fertilizers are divided on two types depending on the fungus itself; there are bio-fertilizers based on ectomycorrhizal fungi and bio-fertilizers based on arbuscular mycorrhizal fungi. Arbuscular mycorrhizal fungi (AMF) promote a significant increase of the area of root absorption of plants colonized, maximizing the use of water and nutrients [8, 9]. These symbiotic fungi enhance plant resistance to water stress, to high temperatures, improve resistance to conditions of toxicity and acidity of soil and to divers' type of pathogens [8]. In addition to soil stabilization in the form of aggregates [9]. AMF are known for their positive effects on phosphorus assimilation by the mineralization of organic phosphorus and solubilization of insoluble phosphorus [10]. In spite of their potential and benefits, the large-scale use of AMFs is still restricted, mainly due to the deficiency of availability of inoculant in high quantities, low cost and high quality, besides the lack of practicality of inoculation in the field [9, 11]. Their efficiency is also questioned by some authors [12, 13], who claim that AMF indigenous community promotes greater root colonization than the addition of commercial inoculants. In this context, the present study, aims to lift the veil on bio-fertilizers based on arbuscular mycorrhizal fungi to ameliorate agriculture in arid lands, their agro-ecological roles, technic of production and the challenges of possibility of installing a bio-fertilizer production unit in these areas.

2. Importance of arbuscular mycorrhizal fungi bio-fertilizers

Arbuscular mycorrhizas are the most common underground mutualistic symbiosis relationship [8]. They are considered as obligate biotrophic organisms that live in the metabolically active roots of terrestrial vascular plants, epiphytes, rhizoids and stems of bryophytes [1–8, 14]. Studies showed that AMF exists 460 million years before first plants originated [2, 15]. They form a mutualistic symbiosis between AMF, belonging to the Glomeromycotina sub-phylum, and 80%

AMF species	Plant	Bio-fertilizers Nature	Stress type	Mechanism used	References
Glomus spp.	Tomato *Solanum hypersicum*	Commercial formulation	Drought	Improving water and nutrient absorption	Kuswandi and Sugiyarto [16]
Rhizophagus intraradices, *Funneliformis mosseae*	Orchard grass (*Dactylis glomerata*)			Improved water content	Kyriazopoulos et al. [17]
Glomus deserticola	*Ocimum basilicum*	Commercial formulation	Soil Salinity	Plant enhancement and alleviation of soil salinity	ElHindi et al. [18]
Mix of height AMF species originating from saline soils	*Tamarix articulata*	Indigenous formulation		Improving plant biomasses, water and nutrient absorption	Bencherif et al. [13]
Rhizophagus intraradices	Black locust (*Robinia pseudoacacia*)	Commercial formulation	Heavy metals pollution	Improved plant biomass causing positive impact on photosynthesis and macronutrient acquisition	Yong et al. [19]
Septoglomus constrictum, *Claroideoglomus lamellosum*, *F. geosporum*, and *F. mosseae*	Alfalfa (*Medicago sativa* L.) and tall fescue (*Festuca arundinacea* Schreb	Indigenous inoculum	Dioxin/furan polluted soils	Improvement of plant dry weight, bacterial, archaeal OUT 's and bacterial diversity	Meglouli et al. [20]
Rhizophagus irregularis, Funneliformis mosseae	*Triticum aestivum*. *Orvantis* and *Lord*	Commercial strain multiplied Commercial formulation	Biotic stress: Oïdium *Blumeria graminis*	Protection against pathogen and reduction of infection	Mustapha et al. [21]

Table 1.
Impact of AMF bio-fertilizers to alleviate biotic and abiotic stress.

of land plant species [8]. AMF are endophytic fungi with intra-radical hyphae that penetrate inside cortical cell and/or the root epidermis [7]. During this fungi and plant interaction, dialogues at molecular level take place, which result in host's metabolic modifications, protection against environmental stresses, and providing friendly conditions to symbiont (fungi) [2]. Arbuscules are formed by endomycorrhizal hyphae within the plant cortical cells and are highly branched, and mature arbuscules have short life and survive for 4–5 days. Arbuscules are considered as functional site for nutrient exchange [8]. These fungi benefit their hosts by increasing the uptake of nutrient elements (especially P) and enhancing the resistance to biotic or abiotic stresses [1, 7]. So far, a large number of publications demonstrated that AMF are an important regulator of the plants performance in different stressed environment (**Table 1**) such as heavy metal contained soils [20, 22], saline soils [13–15, 18, 20, 22, 23], soil subjected to drought stress [24]. AMF are also identified as regulator of biotic stress [21]. AMF association with plants not only improves plant growth but also improves soil texture by changing soil particles into stable aggregates that ultimately resists against wind and water erosion [15]. Mycorrhizal association is also helpful to introduce new plant species in new areas, occurring them ability to survive in a new environmental conditions [23]. Moreover, colonization of root by AMF can arise three sources of inoculum: spores, infected root and hyphae, collectively termed propagules [8].

As well as, with the emergence of sustainable development context the application of AMF bio-fertilizers starts to widen. It is assumed therefore, that the judicious use of these natural inoculants can reduce the need to amend soil with chemical fertilizers, thus increasing the viability of sustainable agriculture [15, 18]. **Figure 1** illustrates the role played by arbuscular mycorrhizal fungi in plant life cycle.

Thus selection of inoculum source constitutes an important parameter for the plantation successful. In fact, the response of AMF to abiotic stresses divers on the level of fungal species or ecotype; it was proven that indigenous strain occurs more beneficial effect into mitigates divers abiotic stress such as water deficiency, saline stress and heavy metals conditions by enhancing the active absorptive surface area which ultimately stimulates the uptake of water and nutrients [13–15, 20, 22, 23].

Figure 1.
Role and importance of AMF symbiosis in plant life.

3. Production of arbuscular mycorrhizal fungi bio-fertilizers

The commercial history of bio-fertilizers began with the launch of 'Nitragin' by Nobbe and Hiltner, a laboratory culture of Rhizobia in 1895, followed by the discovery of Azotobacter and then the blue green algae (BGA) and a host of other micro-organisms. Azospirillum and Arbuscular mycorrhizal fungi are fairly new discoveries [6]. Industrial manufacturing of AMF as crop inoculants is relatively new and, despite the practical demonstrations of their efficiency, their adoption by crop producers has been slow, most likely due to the quality and efficiency of marketed products [11]. In fact, production of previously selected AMF for their use as bio-fertilizers began in the decade of 1990 in a large part of the world [25]. Few companies throughout the world have manufactured and commercialized AMF inoculant using either a single AMF species or mixtures of AMF species that may include plant growth promoting rhizobacteria (PGPR) or other symbiotic and/or biocontrol fungi [11].

The production of inoculum differs from fungi family to another. Arbuscular mycorrhizal fungi are strict obligatory symbionts, they dependent on the presence of a host plant to accomplish their development and their multiplication. The inoculum producer is then required to co-cultivate the "fungus-host plant" complex. Without the use of host plants it would be impossible to complete the mycorrhizal life cycle until the production of new propagules/spores [7]. In addition to this monosporic inoculant, it is possible to produce inoculant with different native species with greater ease and speed [7]. In comparison with the commercial inoculant, it has a low cost, higher taxonomic diversity, and the use of locally adapted species [25, 26], which increases the chances of positive effects on the plant and avoid the introduction of exotic species [27]. The use of AMF inoculant produced from the forest soil is the most reliable and recommended method because of its high species diversity, the potential to accelerate the ecological restoration of the soil environment and to promote the germination and growth of the plants [9, 25–27].

3.1 Conventional method of AMF bio-fertilizers production

This method consists on AMF multiplication on pot culture with selective host plant under controlled conditions in a greenhouse or in a grow room [7].

3.1.1 Mixture of AMF species bio-fertilizers production

Native soil sampled from different plots of same natural sites must be mixed together to create one composite sample. The obtained mixture was distributed into pots (500 mL disposable cups) which were sown with trapping plant aiming to multiply and restore infective structures of the AMF species present in the trap cultures [27], and then kept in a greenhouse for four month. Two plant species are commonly used for trapping culture: cover "*Trifolum repens*" and leek "*Allium porrum*" [7, 27], But use of other legumes is also permitted such as Alfalfa "*Medicago sativa* [13] *Brachiaria* sp. [27]. Once the four month over, the areal part of plants is catted and the soils are mixed with roots for preparing a new plantation for other four months. Simultaneously, at each month, one pot of each plot was taken for analysis, using 50 g of soil for AMF spore isolation and identification and the roots for evaluation of the mycorrhizal colonization rate [26, 27]. A minimum of 12 months is required to obtain a good product, but the ideal is 24 months [13, 20]. Therefore, the obtained inoculums consist of different types of propagules: spores, fungal mycelium and fragments of mycorrhizal roots [7]. Multi-species products are closer to natural conditions because in ecosystems it is rare to encounter only one species of mycorrhizal fungus. The presence of several fungal species allows the inoculum to respond to a greater diversity of culture conditions.

3.1.2 Monospecific AMF species bio-fertilizers production

Production of monospecific AMF bio-fertilizer is based on the use of one selective AMF spore species isolated from natural soil using wet saving method [28], and cultivated on trapping culture with appropriate plant species on pot culture of 15 cm. Three months are needs to obtain AMF multiplication. Verification of AMF sporulation must be done each 20 days to one month. At same sanitary tests can also be performed to ensure that no contamination by parasitic fungi or sporulation of other AMF species has occurred. After four months, monospecific spores are ready for inoculation on seedlings of desired crops. In fact, [29] reported that Rhodes grass (*Chloris gayana*) is the best host for mass multiplication of *Glomus fasciculatum*. Bahia grass (*Paspalum notatum*) was used for multiplication of *Glomus deserticola* [27].

3.2 In vitro technic of AMF bio-fertilizers production

In vitro technic is an aseptic multiplication of AMF on roots cultivated on synthetic medium under sterile conditions. However, this technic started with the early work of [30], and subsequent development by [31, 32] and just after by [33], these authors developed the monoxenic cultivation system to produce contaminant-free AMF, allowing the realization of large-scale production under strictly controlled conditions [34]. The In vitro production of AMF bio-fertilizers consist on the extraction of potential viable propagules from soils with surface sterilization before optimization of growth conditions for germination under aseptic conditions. This aseptic technic consists on cultivation of number of AMF species in association with transformed host roots on synthetic growth medium [33]. Chabot et al. [35] established cultures from surface sterilized spores as starter material and produced 750 spores in 30 ml medium after a period of 4 months of growth in a mono-compartmental petri plate system. This is followed by the association of the propagules with a suitable excised host root for propagule production and recovery. Another system of in vitro AMF multiplication was developed by St Arnaud et al. [36], they used a bi-compartmental Petri plate system and obtained 15,000 spores in 3–4 months. Douds [37] improved this bi-compartmental system by replacing the medium in the distal compartment by fresh medium at regular intervals and obtained 65,000 spores in the distal side of the bi-compartment in a period of 7 months [34]. This technic of bi-compartment petri plate permitted to produce more than 250,000 propagules in 10 ml of medium, which made this technology attractive for industry. However, many process controls must be done to reduce the level of contamination, what should not exceed 3–5% [7, 34]. Once the AMF product is obtained; mass-produced propagules are then formulated in an utilizable form and stored before application to the target plant [7–34, 37]. This technic facilitates the efficient utilization of space and energy in the production system, using solid-state fermentation. Since the technology is more dependent on personnel, it lowers the number of man-days and achieves higher productivity [34].

The use of this technology remains useful for in vitro laboratory tests, but the mycorrhizal inoculum thus obtained (artificial environment on genetically modified roots) is not suited to applications in the agricultural field, providing overall unsatisfactory results [4].

3.3 Production of arbuscular mycorrhizal fungi bio-fertilizers in arid lands

Both conventional method and in vitro methods are practiced in arid areas to produce AMF bio-fertilizers. Several researches was focused on the increasing of plant production in arid land using AMF inoculum (**Table 2**), conventional method with AMF mixture was the most important technic of production adopted. Nevertheless, in vitro technic was also practiced such as by the energy and resource institute of India (TERI) [34]. They based on the faculty of Glomus genus to provide the possibility of using colonized roots as inoculum material with up to 80% of root colonization attained at 4 and 12 weeks [34, 39]. Despite, arid lands are often localized in underdeveloped country with low economical budget who cannot afford to allot enormous amounts in order to produce bio-fertilizers, so the conventional method remains the most appropriate technique under these conditions. In addition ecological conditions of arid lands give them specific characteristics that are not accommodating with all AMF strain. For that production of native AMF bio-fertilizers adapted to local conditions and to specific abiotic stress is essential [13]. Labidi et al. [39] developed a native AMF bio-fertilizer adapted to calcareous arid Tunisians soils. Abdelsalam et al. [38] produced AMF inoculum of desert saoudian areas using *Sorghum halepense* as trapping plant. Bencherif et al. [13] developed a specific AMF bio-fertilizer for arid saline soils. It is noted that in

Inoculum production technic	Specific abiotic stress/zone	Propagule richness	Infection level	References
Conventional method	Drought stress/Saoudian areas	20 g of Sudan grass rhizosphere with 950mycorrhizal spores and 0.5 g of colonized roots	78,5%	Abdelsalam et al. [38]
Conventional method	Calcareous/Tunisian areas	*Septoglomus constrictum*, *Funneliformis geosporum*, *Glomus fuegianum*, *Rhizophagus irregularis* et *Glomus* sp	90%	Labidi et al. [39]
In vitro method	Draught/Tafilalet Morocco	Rhizophagus irregularis	100%	Meddich et al. [5]
Conventional method	Draught/ Tafilalet- Morocco	Glomus sp., Sclerocystis sp., Acaulospora sp	15, 9, 1, spores, /gr of soil	Meddich et al. [5]
Conventional method	Heavy metal Polluted soil/Oran- Algeria	Acaulospora sp., Archaeospora sp., Glomus sp., Claroideoglomus sp., Ambispora sp., Diversispora sp	<50%, <20%, <5%,<5%, <5%/g of soil	Sidhoum and Fortas [40]
On-farm method	Drought	*Rhizophagus clarus* and *Claroideoglomus etunicatum*	80%	Moreira et al. [26]

Table 2.
AMF bio-fertilizers produced in arid lands.

all case, efficiency of AMF bio-fertilizers is related to the better combination AMF genotype/host plant genotype/adaptation to specific abiotic stress [13, 27].

3.3.1 On-farm method

To have AMF bio-fertilizer produced at a large scale with low cost, studies has been developed to test the multiplication of AMF spores under field conditions, called the "on-farm method". These studies explore AMF colonization with strains isolates that are environmentally adapted to native environmental conditions, which potentially representing a low-cost alternative for farmers [26, 37]. This technic is based on sowing plants seeds or seedlings in intact soil cores or mixed soil samples for sufficient time to allow development of AMF symbiosis and then roots were sampled, processed and assessed to measure mycorrhiza formation [26]. Indeed, [41] showed that multiplication on-farm with *Rhizophagus clarus* and *Claroideoglomus etunicatum* grown in agro-industrial residues, such as sugarcane bagasse, is a good strategy for the multiplication of AMF, leading to excellent inoculum potential and large numbers of spores. As well as, the on-farm technic allows farmers and nursery workers to access inoculums with the most effective AMF strains for their culture and their soil and climate conditions; furthermore, they can produce seedlings already inoculated with adapted AMF strains, which enhancing their establishment of s in the field conditions. In addition Moreira et al. [26] produced AMF inoculum with *Rhizophagus clarus*, *Claroideoglomus etunicatum* species, and native AMF from pineapple and coffee plantations, using spores multiplied by the on-farm method to enhance the growth of pineapple and coffee plantlets. These authors concluded that AMF inoculum favorite growth of the commercial tested crops with a high viability of AMF spores. This method is recommended for arid land due to their specific AMF strain and low coast, it could be applied and generalized in order to developing agricultural practices in these areas. Furthermore, because fungi carry different amounts of nutrients for plants, they may affect the growth of plants differently [37], it is preferable to use mixed AMF strains adapted to native conditions which could maximize the absorption of limiting nutrients [26]. This phenomenon could provide more benefits compared to colonization with exotic AMF strains or with single AMF species. Moreover, the mixed inoculation of AMF might have the characteristic of complementarity, exploiting the best of each species that colonizes the plant [26–37, 41].

3.4 Formulation of AMF bio-fertilizers

Formulation technologies largely take care of possible adverse environmental effects and factors that may render the inoculum ineffective [34]. In fact, bio-fertilizers are generally prepared as liquid suspension or more generally solid support containing different types of propagules: spores, fungal mycelium, mycorrhizal root fragments [7]. Generally AMF bio-fertilizers are presented with multi AMF species, which are closer to natural conditions; because in natural ecosystems it is rare to encounter only one species of mycorrhizal fungus. The presence of several fungal species allows the inoculum to respond to a greater diversity of culture conditions [13, 39]. Bio-fertilizer support can be composed of peat, vermiculite, lignite powder, clay, talc, rice bran, granulated rock phosphate, charcoal, soil, straw compost of rice or wheat otherwise a mixture of these materials. In current practice, for better framework life of the bio-fertilizer formulation, the support is selected on the basis of the viability of the microorganisms mixed with them. Likewise, the pre-sterilization of the support and its nutrient enrichment is the other strategy to improve framework, allowing AMF to be maintained in a non-competitive microenvironment. Sucrose, maltose, trehalose, molasses, glucose, and glycerol are additional nutrients and cell-protecting agents commonly used along with a support to ensure maximum cell viability and extended shelf life [4, 39]. After production, AMF bio-fertilizer may be in the form of granules, powder, tablets, pralins, or liquid suspension:

- The micro granules, between 1 and 4 mm, are easily mixed with support of mycorrhizal plants or brought into the planting surface as close as possible to the roots;

- Very fine powders (particles <250 μm) make it possible to prepare a suspension which can be sprayed on growing media or injected into the soil at the base of plants already installed. This type of inoculum can also be used as a seed coating.

- The tablets allow easy dosing of the inoculum to be placed in the plantation area. The inoculum provided is localized in one place and not distributed evenly over the roots.

- Associated with a praline, the inoculum is particularly suitable for plants with bare roots. In a single operation, the plant is inoculated and its roots protected.

- Liquid suspensions are suitable for coating seeds. These inocula can also be sprayed on growing media or injected into the soil at the base of plants [4–7, 39].

The formulated AMF bio-fertilizers should be positioned near the roots, with avoiding the spreading products and favor the injection or burying method. Finally, it is also possible to produce "2 in 1" products:

- Inoculated planting support, ready to use and particularly suitable for soil-less culture,

    - Pre-inoculated plants (vine, chestnut, etc.) ready for planting and whose mycorrhization has been checked before marketing,

    - Seeds coated with AMF propagules (mainly spores) which make it possible to spread and inoculate a plot in a single pass,

    - Organic fertilizers and amendments containing AMF propagules [7].

4. Design of AMF bio-fertilizer production unit

The aim of our present study is to set up a production unit for AMF bio-fertilizers, which has the advantage of covering the biological deficit of arid soils, a problem which continues to degenerate more and more. Indeed, the installation of a bio-fertilizer unit must follow some criteria such as: appropriate production, location, construction space, equipment, machinery, other laboratory equipment and working capital. The bio-fertilizer production unit should be founded in a homogeneous area based on the interaction of soil characteristics, geomorphology and climate. This place should be qualified as a buffer to reduce the risk of contamination during the process of production and quality control. The overall architecture of the unit is the key element for the success of any economic project. In our case, the bio-fertilizer production unit must be organized in an H-shaped architecture (**Figure 2**). This comes, according to Alamari [4], This author explained that this structure is based on the fact of its economical aspect and its ability to adapt to the sterilization process with forward walking, in other words from the "soiled", to go towards the "clean" and then towards the "sterile", without possibility of going back and without crossing of flows of "soiled" and "clean". The unit must contain administration, laboratories, storage and packing space.

4.1 Administration

The administrative team must ensure the respect of directives and guidelines in addition to external relations allowing the best conditions for the scientific team to carry out their work. To achieve these objectives, different tasks must be implemented: (i) Preparation, application and management of the budget and control of its execution; (ii) Establishment of contracts and agreements with different organizations in the same field, while taking care of calls for submissions and the

Figure 2.
Conventional method for production of mixed AMF bio-fertilizers

various procedures for procuring equipment; (iii) Physical and telephonic reception is provided; in addition to management of the staff and personnel of the unity by application and monitoring of health and safety guidelines.

4.2 Laboratories

The setting up of a laboratory should meet the criteria approved by the World Health Organization (WHO). A laboratory should be built with walls, ceilings and horizontal surfaces, non-polished, easy to clean, impermeable to liquids and resistant to disinfectants and to antiseptics. Then, to ensure the best work conditions, the laboratory areas must be spacious [42]. In addition, the laboratory must contain:

- Mechanical ventilation system ensuring interior air movement without recycling;

- Electricity must be sufficiently powerful with an emergency restart system in the event of blackouts;

- The town gas supply must be appropriate and protected.

- Presence of the cleaning tank with emergency shower next to each laboratory door, with installation of physical protection and fire safety systems.

- Finally, providing enough materiel resources and space for treatment and safe storage of solvents, radioactive substances as well as compressed and liquefied gases.

The production unit laboratory of AMF bio-fertilizer must contain three compartments: (i) Greenhouse for AMF inoculum multiplication; (ii) In vitro multiplication and strains isolation rooms; and (iii) compartment of control, conditioning and storage.

4.2.1 Greenhouse AMF multiplication

The greenhouse for AMF multiplication must be positioned behind the unit occupying a clear space with 204 m^2 of approximately area. The greenhouse is used to care for young host plants and to maximize crop productivity by improving the relationship between their growth and AMF biomass. Greenhouse is the most important compartment of the unit, so the geographical location may have to be considered. The attractive location must be related to the adaptability and value of the land, cost of fuel delivered, ample and inexpensive water, in response of number of question: (i) What is the yearly available solar energy? (ii) How much moisture falls, summer and winter? (iii) What are the maximum and minimum temperatures and their duration? (vi) What are the hail and wind belts?(v) Is air pollution a potential problem? (vi) Information on all of the foregoing questions allows the greenhouse operator to determine the degree to which he can maintain near optimum environmental conditions for plant growth and AMF multiplication [43]. Wind is important climatic problem in arid lands, so the wind direction plays an important role in the choice of unit implantation site. So, orientation of the greenhouse is a compromise for wind direction, latitude of location and type of temperature control [44]. After site choice, the greenhouse must follow some recommendations. The greenhouse must be constructed in glass farmed structure on double-sloped with a naturally exposition of natural light for much of the day.

In addition, heat is partly assured by sun rays paired with artificial means, such as circulating steam, hot water, or hot air. Ventilating system is also needed [43, 44]. For low coast, the ventilation must be assured by roof openings and large windows on the side, which can functioned mechanically or automatically. In some conditions, if a financial condition allows automatic ventilation system is installed. For the AMF multiplication, trapping culture must be done in pots or in specialized containers. In this setting, trapping cultures are grown for 3 to 4 months to minimize the accumulation of saprophytes in the medium for excessive growth and senescence of the roots. However, culture maintained for more than 5 months and regulated watering is recommended before areal part cutting and replantation of new seeds as explained above (3.1.1).

4.2.2 In vitro multiplication room

It represent an aseptic areas separated from the greenhouse and the AMF isolation room in order to avoid any contamination and to control sterility conditions as much as possible.

4.2.3 Drying and conditioning room

This area is located just before the greenhouse; it is used to dry the contents of the pots and containers for later conditioning. Once the trapping plants are ready to be harvested, they are moved to shelves in this area so that they are not exposed to light. Drying take about 2 to 3 weeks. After this period, the roots of the trapping plants are cut and mixed with a suitable substrate. Conditioning AMF inocula begins by placing the cultures in sealed bags. These bags are provided with codes written both on the surface and on labels affixed to the upper left corner of the crops so that they are easily identified when stacking. In addition, an organization in alphabetical order of cultures is also recommended.

4.2.4 AMF strain isolation room

Isolation room is completely isolated from all plant growing areas and the use of unsterilized soil is strictly prohibited. Isolation is practiced as follows: The contents of the dried pots are installed on grounds. The isolation of the spores from the sample is done by wet sieving method [28]. This technic of isolation is practiced in order to produce bio-fertilizer containing AMF spores. In addition isolated AMF strains are conserved in order to develop further research. During the AMF isolation process, a series of precautions must be observed, especially disinfection of surface area of isolation, tables and shelves with draying after cleaning. Asepsis is main condition for the success of this crucial stage of the AMF bio-fertilizer production.

4.2.5 Control room

This space is used for carrying a series of bio-fertilizer control tests. These tests include AMF spores count with microscopic examination, evaluation of AMF root colonization rate and elaboration of must probable number test. AMF spores number must vary between 10 and 15 pots per day. Once extracted, the spores are transported in glass Petri dishes and stored in the laboratory refrigerator. Indeed, the examination is carried out by a stereo microscope on the day of the extraction. The information thus retrieved is stored in a database and all written notes are archived as a physical backup. Systematically, all the files having processed the culture collection are stored centrally on the unit's web server, saved on a separate hard drive on the same computer, and stored on another computer in the laboratory.

4.2.6 Storage room

The storage is done at an atmosphere of 4°C at the level of the shelves of metal racks. These are characterized by mesh surfaces to optimize air circulation and facilitating their cleaning. The racks are placed in the center of the room and equipped with wheels to facilitate their movement. The storage period can reach a maximum of 3 years [4]. Bio-fertilizer storage process requires certain recommendations mainly: bag surface and their labels must be cleaned and disinfected before they are placed in this room. Floors and shelves are regularly disinfected. AMF bio-fertilizer product should be stored in a corner of the laboratory where the air temperature is not detrimental to the viability of AMF propagules.

5. AMF bio-fertilizers challenges of production and application on agricultural projects of arid lands

5.1 Why produce native AMF bio-fertilizers for arid lands?

Arid lands constitute the most widespread terrestrial biome in earth, with 35% of the land areas of the world. These areas are subjected to several desertification phenomena [3]. To counteract this problem, applications of new agricultural technics are required including application of bio-fertilizers. Nevertheless, the use of AMF for the restoration of degraded ecosystems has received poor attention, requiring a different approach [25]. In addition, loss of AMF propagules is usually recorded following soil and cover plant degradation, which could further inhibit natural and/or artificial revegetation processes [3, 25]. Taking into account all the previously cited aspects and the necessity of restoration in these areas, the eco-technology proposed by some studies [13–15, 18, 20, 22, 23, 25] represents a good alternative. They propose the restoration of degraded areas by re-introduction of native AMF and plant species [25].

5.2 How produce native AMF bio-fertilizer unit?

Production of native AMF bio-fertilizers unit require appropriate funding with adoption of a good financing strategy, based on various technical-economic parameters including description of the income elements and those of the expenses (**Table 3**). Indeed, the elements of income include the sale of bio-fertilizers, remunerations, publicity and assurance. **Table 4** describes all the expenses and revenues provided by the bio-fertilizer production unit.

5.3 Economical challenge of native AMF bio-fertilizer production?

Production of native AMF production is an agro-industrial investment for each arid region, in some case the product may be exported to similar edapho-climatic areas. Designed production and marketing chain must support each unit in order to guarantee the success of the project. In **Table 4** we have tried to establish an approximate economic study which will allow the investors to have an idea on the economic situation of the project, but these figures may vary from one country

Space	Surface (m^2)	Realization coast ($)
Reception	35.61	91,873,800
Conferences room	31.91	82,327,800
Manager's office	17.3	44,634,000
Engineers office	56.02	144,531,600
Storage room	55.13	13,235,400
Isolation room	40.55	104,619,000
Multiplication room	61.88	524,436,600
Conditioning room	64.6	1,666,668,000
Quality Control room	97.99	252,814,200
AMF greenhouse multiplication	203.27	655,545,750
Sales office	41.63	107,405,400
Store checkout	44.34	114,397,200
Changing room and toilets	73.23	188,933,400
Total	823,46	3,991,422,150

Table 3.
Realization coast of native AMF bio-fertilizers unit.

Expenses ($)			Incomes
Investment costs	Construction field	65,000/m^2 in average	24$/kg of Native AMF bio-fertilizer
	Construction	**4,000,000,000**	
	Technical installation	40,000	
	Materials	100,000	
	Tools	100,000	
	Transport	100,000	
Production charge	Raw material production	13,500,000	
	Packaging	9,030,000	
	Salaries	Varied between 3,225,000 and 10,320,000/employee	
	Repair and maintenance	10%	
	Insurance	10% of the operating budget	
	Publicity	10%	
Supplementary charge		2–3%	
Total	850 milliards of dollars of investment	12900 millions of dollars /year	

Table 4.
Expenses and revenues of native AMF bio-fertilizers unit.

to another. The main thing is to ensuring the success of the plant development, potentiates the in situ conservation of the AMF community and preservation of ecosystem stability and biodiversity (**Figure 3**).

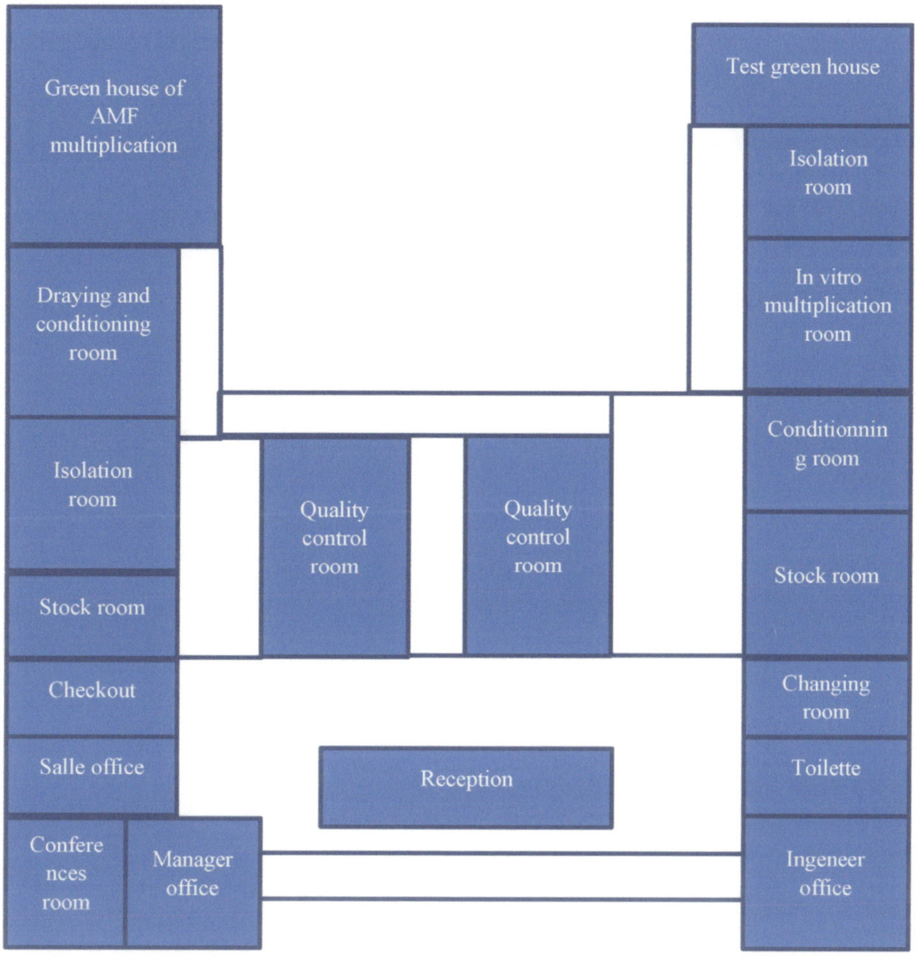

Figure 3.
Representative plan of AMF production unit on H form (Adapted from Alamri [4] description).

In absence of information about the stapes and strategies of native AMF bio-fertilizer unit, these information given in **Tables 3** and **4** can help in understanding and progressing of this strategy and its application in degraded arid lands for development of sustainable agriculture.

6. Conclusion

Arid land has specific ecological characteristics that confer them special management strategies. As same the different soil throughout the world, soil of arid land present a diversity of AMF that may be exploited to enhance agriculture in these areas. For that in the present chapter we have exposed the most important technic of AMF bio-fertilizer production based on native AMF strains. Founded on what has been explained, recommending the use of conventional pot culture technic present the adequate method adapted for arid lands. We have also given an approximate economical evaluation for coast of building and installing native AMF production unit in arid lands. The conventional method includes optimization of scale of production while using the adequate trap plant, law-coast and environmentally safe. With this technology, the preservation of environment from chemical fertilizer pollution must be enhanced, which lead to economic development of agro-industry adapted for each country with operative system for sustainable agriculture.

Acknowledgements

Authors thank every person helps to realize this work.

Conflict of interest

The authors declare no conflict of interest.

Author details

Bencherif Karima* and Therrafi Samia
Faculty of Life and Nature Sciences, University of Djelfa, Djelfa, Algeria

*Address all correspondence to: bencherif_karima@yahoo.fr

References

[1] Sellosse MA. La symbiose: structure et fonction, Rôle écologique et évolutif. Ed Vuibert. 2000:138

[2] Sellosse MA. Jamais seul: ces microbes qui construisent les plantes, les animaux et les civilisations. Ed Actes Sud. 2017:357

[3] Tchakerian VP, Deserts PP, In D. Developments in Earth Surface Processes. Elsevier (eds). 2015;**19**:449-472 https://doi.org/10.1016/B978-0-444-63369-9.00014-8

[4] Alamari R. Installation d'une unité de production de biofertilisants. Mémoire d'ingéniorat en agronomie. Institut national agronomique de Tunisie. 2016:74

[5] Meddich A, Oufdou K, Boutasknit A, Raklami A, Tahiri A, Ben-Laouane R, et al. Baslam M. Use of Organic and Biological Fertilizers as Strategies to Improve Crop Biomass, Yields and Physicochemical Parameters of Soil. In: Meena RS, editor. Nutrient Dynamics for Sustainable Crop Production Springer Nature Singapore Pte Ltd. 2020. p. 247 https://doi.org/10.1007/978-981-13-8660-2_9

[6] Mazid A, Khan TA. Future of Bio-fertilizers in Indian Agriculture: An Overview. International Journal of Agricultural and Food Research. 2014;**3**(3):10-23

[7] Fortin AJ, Plenchette C, Les mycorhizes la nouvelle révolution verts PY. Eds MultiMondes. In: P138. 2008

[8] Smith SE, Read DJ. Mycorrhizal symbiosis. 3rd ed. New York: Academic Press – Elsevier; 2008

[9] Dos Santos RS, Ferreira JS, Scoriza RN. Inoculum production of arbuscular mycorrhizal fungi native to soils under different forest covers. Rev. Ceres. Viçosa. 2017;**64**(2):109-111

[10] Priou L. Multiplication des mycorhizes arbusculaires en milieu liquide et solide afin d'améliorer la formulation de biofertilisants. Sciences agricoles. 2013. ffdumas-00975007.

[11] Gianinazzi S, et Vosatka M. Inoculum of arbuscular mycorrhizal fungi for production systems: science meets business. Canadian Journal of Botany. 2004;**82**:1264-1271

[12] Gianinazzi S. La biotechnologie des mycorhizes à arbuscules en horticulture. In: Alliances au pays des racines –14. Colloque Scientifique de la Société d'Horticulture de France. (25.05.2012). 2012. pp. 14-16

[13] Bencherif K, Dalpè Y. Lounés Hadj-Sahraoui A. Influence of Native Arbuscular Mycorrhizal Fungi and *Pseudomonas fluorescens* on Tamarix Shrubs Under Different Salinity Levels. Soil. Biology. 2019:275-284

[14] Linderman RG. Role of VAM fungi in biocontrol. In: Pfleger FL, Linderman RG, editors. Mycorhizae and plant health. American Phytopathological Society, St. Paul. 1994. pp. 1-25

[15] Nadeem SM, Khan MY, Waqas MR, Binyamin R, Akhtar S, Zahir ZA. Arbuscular Mycorrhizas: An Overview. In: Wu Q-S Editor. Arbuscular Mycorrhizas and Stress Tolerance of Plants. Ed. Springer; 2017. pp. 1-24

[16] Kuswandi PC, Sugiyarto L. Applicaton of mycorriza on planting media of two tomato varieties to increasevegetable productivity in drought condition. Jurnal Sains Dasar. 2015;**4**:17-22

[17] Kyriazopoulos AP, Orfanoudakis M, Abraham EMAEM, Serafidou N. Effects of arbuscular Mycorrhiza Fungi on Growth Characteristics of Dactylis

glomerata L. under Drought Stress Conditions. Notulae Botanicae Horti Agrobotanici Cluj-Napoca. 2014;**42**(1):132-137. DOI: 10.15835/nbha4219411

[18] Elhindi KM, Sharaf eddine A, Elgorban MA. The impact of arbuscular mycorrhizal fungi in mitigating salt-induced adverse effects in sweet basil (*Ocimum basilicum* L.). Saudi Journal of Biological Sciences. 2017;**24**:170-179

[19] Yang Y, Liang Y, Han X, et al. The roles of arbuscular mycorrhizal fungi (AMF) in phytoremediation and tree-herb interactions in Pb contaminated soil. Science Reporter. 2016;**6**:20469

[20] Meglouli H, Lounès-Hadj Sahraoui A, Magnin-Robert M, Tisserant B, Hijri M, Fontaine J. Arbuscular mycorrhizal inoculum sources influence bacterial, archaeal, and fungal communities' structures of historically dioxin/furan-contaminated soil but not the pollutant dissipation rate. Mycorrhiza. 2018 https://doi.org/10.1007/s00572-018-0852-x

[21] Mustafa G, Randoux B, Tisserant B, Fontaine J, Magnin-Robert M, Lounès-Hadj Sahraoui A, et al. Phosphorus supply, arbuscular mycorrhizal fungal species, and plant genotype impact on the protective efficacy of mycorrhizal inoculation against wheat powdery mildew. Mycorrhiza. 2016;**26**:685-697

[22] Lenoir I, Lounes-Hadj Sahraoui A, Laruelle F, Dalpe Y, Fontaine J. Arbuscular mycorrhizal wheat inoculation promotes alkane and polycyclic aromatic hydrocarbon biodegradation: Microcosm experiment on aged-contaminated soil. Environmental Pollution. 2016;**213**:549-560

[23] Bencherif K, Boutekrabt A, Laruelle F, Dalpe Y, Fontaine J, Lounés Hadj-Sahraoui A. Impact of soil salinity on arbuscular mycorrhizal fungi biodiversity and microflora biomass associated with *Tamarix articulata* Vahll rhizosphere in arid and semi-arid Algerian areas. Sciences of the total environment. 2015;**533**:488-494

[24] Wu QS, Zou YN. Arbuscular Mycorrhizal Fungi and Tolerance of Drought Stress in Plants. In: Wu Q-S Editor. Arbuscular Mycorrhizas and Stress Tolerance of Plants. Ed. Springer. 2017. 25-42

[25] Torres-Arias Y, Forsb RO, Nobreb C, Gómeza EF, Berbarab RLL. Production of native arbuscular mycorrhizal fungi inoculum under different environmental conditions. Brazilien journal of microbiology. 2017;**48**:87-94

[26] Moreira BC, Junior PP, Jordão TC. De Cássia Soares da Silva M, Ferreira Ribeiro AP, Stürmer SL, Chamhum Salomão LC, Otoni WC. Megumi Kasuya MC. Effect of Inoculation of Pineapple Plantlets with Arbuscular Mycorrhizal Fungi Obtained from Different Inoculum Sources Multiplied by the On-Farm Method. Rev Bras Cienc Solo. 2019;**43**:e0180148

[27] Sadhana B. Arbuscular Mycorrhizal Fungi (AMF) as a Biofertilizer- a Review. Int.J.Curr.Microbiol.App.Sci. 2014;**3**(4):384-400

[28] Gerdemann JW, Nicolson TH. Spores of mycorrhizalEndogone species extracted from soil by wet-sieving anddecanting. Transactions of the British Mycological Society. 1963;**46**:235-244

[29] Sreenivasa,M.N. Bagyaraj,D.J *Chloris gayana* (Rhodes grass), a better host for the mass production of *Glomus fasciculatum*. Plant Soil. 988;**106**:289-290

[30] Mosse B, Hepper CM. Vesicular arbuscular mycorrhizal infections in root organ cultures. Physiological Plant Pathology. 1975;**5**:215-223

[31] Strullu GD, Romand C. Méthodes d'obtention d'endomycorrhizes à vésicules et arbuscules en conditions axéniques. C R Acad Sci Paris. 1986;**303**:245-250

[32] Strullu GD, Romand C. Culture axénique de vésicules isolées à partir d'endomycorrhizes et réassociation in vitro à des racines de tomate. C R Acad Sci Paris. 1987;**305**:15-19

[33] Bécard G, Fortin JA. Early events of vesicular arbuscular mycorrhizal formation on Ri T-DNA transformed roots. The New Phytologist. 1988;**108**:211-218

[34] Adholeya A, Tiwari P, Singh R. Large-Scale Inoculum Production of Arbuscular Mycorrhizal Fungi on Root Organs and Inoculation Strategies. In: by S, editor. Soil Biology, Volume 4 In Vitro Culture of Mycorrhizas. Declerck: D.-G. Strullu, and A. Fortin). Springer; 2005. pp. 315-338

[35] Chabot S, Becard G, Piche Y. Life cycle of Glomus intraradix in root organ culture. Mycologia. 1992;**84**:315-321

[36] St-Arnaud M, Hamel C, Vimard B, Caron M, Fortin JA. Enhanced hyphal and spore production of the arbuscular mycorrhizal fungus Glomus intraradices in an in vitro system in the absence of host roots. Mycological Research. 1996;**100**:328-332

[37] Douds DD Jr. Increased spore production by Glomus intraradices in the split-plate monoxenic culture systemby repeated harvest, gel replacement, and re-supply of glucose to the mycorrhiza. Mycorrhiza. 2002;**12**:163-167

[38] Abdelsalam E, Alatar A, ElShiekh MA. Inoculation with arbuscular mycorrhizal fungi alleviates harmful effects of drought stress on damask rose. Saudi Journal of Biological Sciences. 2018;**25**:1772-1780

[39] Labidi S, Lounès-Hadj Sahraoui A, Tisserant B, Laruelle F, Rjaibia W, Hamdi K, et al. Intérêt d'un fertilisant mycorhizien local dans la croissance des plantes en grandes cultures. Journée Nationale sur la valorisation des résultats de la Recherche dans le domaine des Grandes Cultures Tunis. In: le 17 avril. 2014

[40] Sidhoum W, Fortas Z. Growth and mycorrhizal responses to cadmium stress in some halophytic plants. Soil Environ. 2018;**37**(2):169-177. DOI: 10.25252/SE/18/61564.

[41] Schlemper and Stürmer, 2014; Schlemper TR, Stürmer SL. On farm production of arbuscular mycorrhizal fungi inoculum using lignocellulosic agrowastes. Mycorrhiza 2014;24:571-580. https://doi.org/10.1007/s00572-014-0576-5

[42] WHO. World Health Organisation. In: Ethical conditions to establish aseptic laboratory. 2018

[43] Hanan JJ, Holley WD, Goldsbery LK. Greenhouse management. Bommer FR et al., editors.1978.Advenced series in agricultural sciences 5. Springer Eds.

[44] Rodríguez F, Berenguel M, Guzmán JL, Ramírez-Arias A. The Greenhouse Dynamical System. In: Modeling and Control of Greenhouse Crop Growth. Advances in Industrial Control. Springer, 2015. Champions. https://doi.org/10.1007/978-3-319-11134-6

Chapter 8

Mycorrhizae Applications in Sustainable Forestry

Dayakar Govindu, Anusha Duvva and Srinivas Podeti

Abstract

Arbuscular mycorrhizal (AM) association is the most common symbiotic association of plants with microbes. AM fungi occur in the majority of natural habitats and they provide a range of important biological services, in particular by improving plant nutrition, abiotic resistance, and soil structure and fertility. AM fungi also interact with most crop varieties and forest plants. The possible benefit of AM fungi in forestry can be achieved through a combination of inoculum methods. The mycorrhizal inoculum levels in the soil and their colonization in different forest plant roots which leads to reduce the fertilizers, pathogen effects and fungicides and to protect topsoil, soil erosion, and water-logging. Currently, several reports were suggested that AM symbiosis can improve the potential for different plant species. Two steps could be used to produce high yielding of different plant biomass that would be both mycorrhizal dependency and suitability for sowing into the field with high inoculum levels Therefore, the wide-scale inoculation of AM fungi on forest trees will become economically important. The successful research is required in the area of mass production of AM fungal inoculum and AM fungi associated with roots which will contribute to sustainable forestry.

Keywords: mycorrhizae, sustainable forestry, mycorrhizal dependency, biomass

1. Introduction

Arbuscular mycorrhizal fungi are ubiquitous soil microorganisms. AM fungi have great potential to increase plant growth and soil aggregate formation which improves soil quality and development of plant health [1] AM fungi and microorganisms in the rhizosphere contaminate roots and fabricate rich nutrients condition for plant development [2]. The advantage of AM growths in the field conditions with singular organisms is commonly identified with the rate and degree of mycorrhiza arrangement [3]. Management of AM fungi is required in forestry for high yielding and biomass production to be derived from economic and environmental conditions. Determining the magnitude of benefits from improved AM fungi management the three important factors are required. 1. Mycorrhizal dependency 2. Nutrient status of soil 3. Potential of AM fungal inoculum. The terrestrial plant roots develop AM fungi with natural resources of 80–90% symbiosis [4]. Global occurrence in forest ecosystems and form 50% of microbial biomass in the tropical ecosystem [5]. Among different important functions of the plant-fungal symbiosis, plant growth promotion activity is stimulated by the phosphorus uptake [6]. AMF increase nutrient uptake for the plants, particularly immobile nutrients such as phosphorus (P), copper (Cu), and zinc (Zn) in the soil which are not accessible to plant roots in normal condition due to slow immobility [7].

Moreover, AM fungi support tolerance to the plants from different environmental stresses such as salinity, drought, heat, and pollutants in the rhizosphere soil [8, 9]. Presently, effective management of AM fungi is possible by agronomic practices.

Mycorrhizal dependency (MD) is the most important in developing the management of crop plants and forest trees. Forest plant species derive profit from AM fungi to facilitate equally, another crop species is highly dependent on AM fungi for nutrition, biomass, and growth [10]. The most agricultural plants and forest tree species are hosts of AMF, not all benefit equally. The RFMD (relative field mycorrhizal dependency) proposed by Plenchette et al. [11] expresses the difference in dry biomass between mycorrhizal and non-mycorrhizal treatments as a percentage of the biomass of mycorrhizal treatment. This method is very useful in the ranking of different host plants with an individual experiment but absolutely the values of RFMD will depend on the nutrient status of the soil. Abbot and Robson [12] have suggested the need to assess the importance of AM to a host across a full range of soil P levels by determining the response curves for mycorrhizal and non-mycorrhizal plants.

In some forest soils, the response of some crop species to AM fungi is expressed as N as well as phosphate benefits [13]. AM inoculation did not significantly increase shoot dry matter of rice, but it produced significantly higher in biomass than the non-mycorrhizal ones.

A perusal of the **Table 1** reveals that all the 10 tree saplings have shown mycorrhizal infection. However, the percent of colonization varied with the tree species. Maximum colonization was observed in *Azadirachta indica* followed by *Albizia lebbeck, Gliricidia maculate.* Least colonization was observed in *Tamarindus indica.* The differences in infection are due to edaphic conditions and the age of the plant. With few exceptions, a direct correlation can be observed between the percent of mycorrhizal colonization and phosphorus content of the plants. Although the saplings with a high percent of colonization show high MD, there is no direct correlation between these two parameters. For instance, the *Leucaena leucocephala* with 78 colonization has shown more MD than *Azadirachta indica* with the highest AM fungal colonization. Thus it is obvious

S. No.	Plant species	% of colonization	P content (%)				Mycorrhizal Dependency (MD)
			Mycorrhizal		Non-mycorrhizal		
			Shoot	Root	Shoot	Root	
1	*Acacia nilotica*	67	0.20	0.30	0.12	0.20	170
2	*Albizia lebbeck*	88	0.80	0.90	0.60	0.70	240
3	*Albizia procera*	72	0.70	0.90	0.50	0.60	210
4	*Hardiwikia binata*	76	0.50	0.70	0.23	0.28	196
5	*Gliricidia macula*	80	0.70	0.80	0.30	0.40	215
6	*Leucaena leucocephala*	78	0.60	0.80	0.31	0.70	213
7	*Acacia melanoxylon*	71	0.40	0.40	0.20	0.26	183
8	*Azadirachta indica*	90	0.90	0.90	0.71	0.80	253
9	*Tamarindus indica*	54	0.10	0.10	0.09	0.12	104
10	*Tectona grandis*	78	0.60	0.80	0.70	0.83	212

Table 1.
Mycorrhizal dependency of some foresty tree species saplings.

that the extent of mycorrhizal colonization has no relation with MD. The plants even with moderate infection may also exhibit high Mycorrhizal dependency. Mycorrhizal dependency of *Acacia nilotica* lowers as the P-level in soil was increased [14].

2. Distribution of AM fungi in forestry

Forests play a progressively more crucial role in gathering the demand for timber and ecological protection, nearly 25% (one fourth) of India's total land area is now under forest land and tree cover. The diminishing soil quality is the main warning to sustainable forest management, mostly in planted forests. Microorganisms show significant functions in soil formation, aggregation nutrient cycling, nutrient uptake, and reclamation of ecosystems [15]. The arbuscular mycorrhizal fungi (AMF) form symbiotic associations by the plant roots of more than 80% of plants [16], and they play a crucial role in the successful organization and maintenance of plant communities [17]. AMF hyphae can add phosphorous (P), which cannot be absorbed by root hairs, and the AMF soil mycelial arrangement provides many profits to host plants [18] as well as plant growth promotion [19, 20] and development of plant resistance to abiotic stress and disease [21]. Additionally, AM fungi can be favorable to soil aggregation as the outcome of the activities of hyphae and glomalin protein secretion [22] therefore, the incidence and colonization of AM fungi would be helpful to the survival of forest seedlings and the sustainable managing of forests. Furthermore, the AM fungal species associated with plant species have elucidated different functions to host plants and influences on the distribution, diversity, and restoration of plant community [23]. The diversity of AM fungi is significantly important to forest ecosystems and can be important for plant community and productivity [24, 25]. Though, information about the diversity of AMF associated with tree species in forest plants is inadequate. It is a known fact the AM fungi are extensive in different ecosystems, and their colonization and spore propagules are also affected by soil physicochemical characteristics [26]. The abiotic factors could influence on root colonization and fungal spore population.

The AM fungi colonization and spore commune compositions in the rhizosphere of the tree species were estimated. The outcome of this study would provide close on the utilization and supervision of AM fungi to keep sustainable management of forests [27].

2.1 Forest soils

Table 2 reveals that AM fungal spore population was found highest in Kothagudem soil of *Albizia lebbeck*, while it was lowest in Godavarikhani soils of *Acacia nilotica*. The AM fungal spore population range was varying in the rhizosphere soils of Kothagudem followed by Yellendu, Bhopalpally, and Kothagudem. On other hand, rhizosphere soil showed a great variation in the incidence of different AM fungi both qualitatively and quantitatively. *Glomus* species was dominating in all the rhizosphere soils, followed by *Gigaspora* species was recorded highest in the rhizosphere of *Acacia nilotica* of Bhopalpally soil and it was least in Kothagudem. Similarly, *Sclerocystis* species was found highest in Kothagudem soil of *Acacia nilotica*, while it was least in Godavarikhani soil of *Acacia nilotica*. *Acaulospora* species was recorded highest in Bhupalpally and it was lowest in Godavarikhani soil. No *Acaulospora* species was observed in Godavarikhani and Yellandu the soils of *Albizia lebbeck*. *Scutellospora* was least in Godavarikhani and Bhupalpally soils. In the rhizosphere soils of the analyzed tree species, bountiful spore numbers, and high decent varieties of AMF species were found (**Figure 1**) [28].

Location	Plant species	Cumulative spore number	Individual spore incidence				
			Glomus	*Gigaspora*	*Sclerocystis*	*Aculospora*	*Scutellospora*
Kothagudem	*A.lebbeck*	155.0 ± 1.53	98	21	32	4	—
	A.nilotica	92.7 ± 1.45	63	12	11	6	—
Bhupalpally	*A.lebbeck*	106.7 ± 1.20	76	12	9	6	3
	A.nilotica	137.0 ± 1.73	72	35	16	9	5
Godavarikhani	*A.lebbeck*	82.3 ± 0.88	48	16	12	—	6
	A.nilotica	63.7 ± 1.45	39	13	8	3	—
Yellandu	*A.lebbeck*	98.2 ± 0.33	57	23	18	—	—
	A.nilotica	118.7 ± 1.45	80	15	10	7	6

Table 2.
Incidence of AM fungi in two Agroforestry tree species of four forest sites of North Telangana. Forest soils.

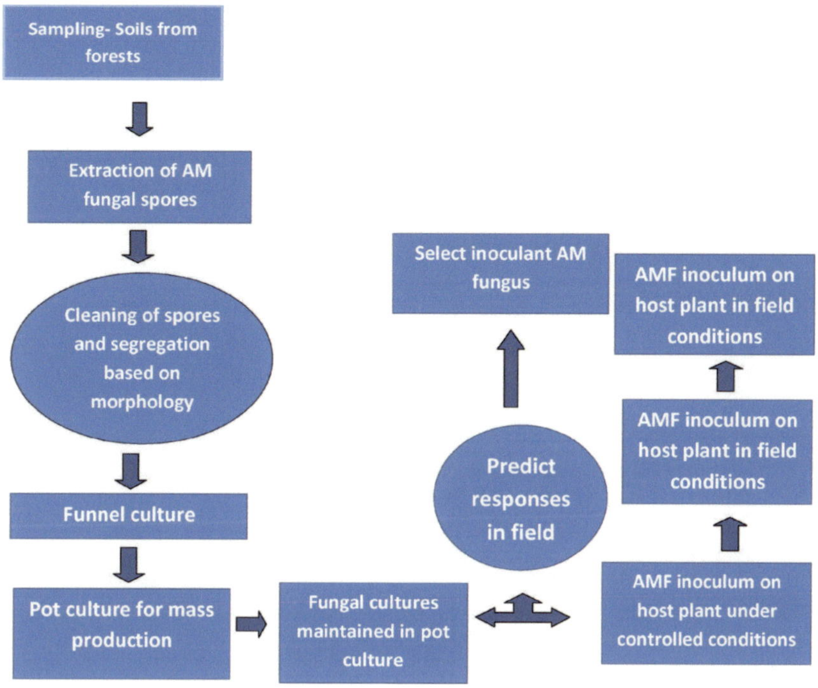

Figure 1.
Isolation and selection of AM fungi for host plant.

2.2 Coal mine soils

Table 3 results showed that AM spore number in rhizosphere soil was studied in different Coal mine sites of North Telangana. The results are depicted that AM spore population was observed in two different plant species. AMF root colonization is presented in **Figure 2**, highest root colonization was recorded in the Kothagudem sample of *Albizia lebbeck*, while it was a significantly higher level of AM fungal spore population was also seen in the same sample. The lowest colonization was found in *Acacia nilotica* of Godavarikhani, but the moderate spore population was observed. The lowest level of mycorrhizal colonization was found in Godavarikhani and Bhopallpally rhizosphere samples of *Albizia lebbeck* and *Acacia nilotica* respectively. However, the AM fungal spore population incidence varies. The highest AM fungal spore population was recorded in Kothagudem soil followed by Bhopalpally and the same trend was observed in Godavarikhani and Yellandu. The AM fungal spore population varied from species to species. *Glomus* species was dominated in all the rhizosphere soil samples of two plants. Past investigations have additionally detailed *Glomus* and *Acaulospora* to be the dominant genera in different woods [26, 29]. *Gigaspora* species was highest in the rhizosphere of *Albizia lebbeck* of Kothagudem, while it was significantly lowest in Yellandu soils followed by *Sclerocystis, Acaulospora,* and *Scutellospora*. Similarly, *Acaulospora* species was found highest in the rhizosphere of *Albizia lebbeck* of Bhopalpally. *Sclerocystis* species was found highest in Kothagudem soils, while it was lowest in Bhopalpally soils. *Scutellospora* spore incidence was observed more in the rhizosphere of *Acacia nilotica* than *A.lebbeck*, but it was found less number in other samples. Interestingly, No *Scutellospora* species was recorded in the rhizosphere of *Albizia lebbeck* in Godavarikhani soil as shown in **Table 4**.

Some AM fungi react diversely to soil disturbances, for instance, Hart and Reader [30] identified that species from the suborder Glomineae were substantially

Location	Plant species	Cumulative spore number	Individual spore incidence				
			Glomus	*Gigaspora*	*Sclerocystis*	*Aculospora*	*Scutellospora*
Kothagudem OCS	*A.lebbeck*	79.7 ± 1.45	31	26	11	8	3
	A.nilotica	61.3 ± 1.76	26	15	9	6	4
Bhupalpally OCS	*A.lebbeck*	55.7 ± 2.33	33	11	4	5	2
	A.nilotica	63.3 ± 0.88	29	12	10	9	3
Godavarikhani OCS	*A.lebbeck*	73.0 ± 1.73	44	14	10	5	—
	A.nilotica	48.7 ± 1.20	17	15	6	6	4
Yellandu OCS	*A.lebbeck*	53.7 ± 1.45	29	10	7	5	2
	A.nilotica	44.3 ± 0.88	15	13	10	4	2

Table 3.
Incidence of AM Fungi in two Agroforestry tree species of four coal mine Opencast sites(OCS) of North Telangana.

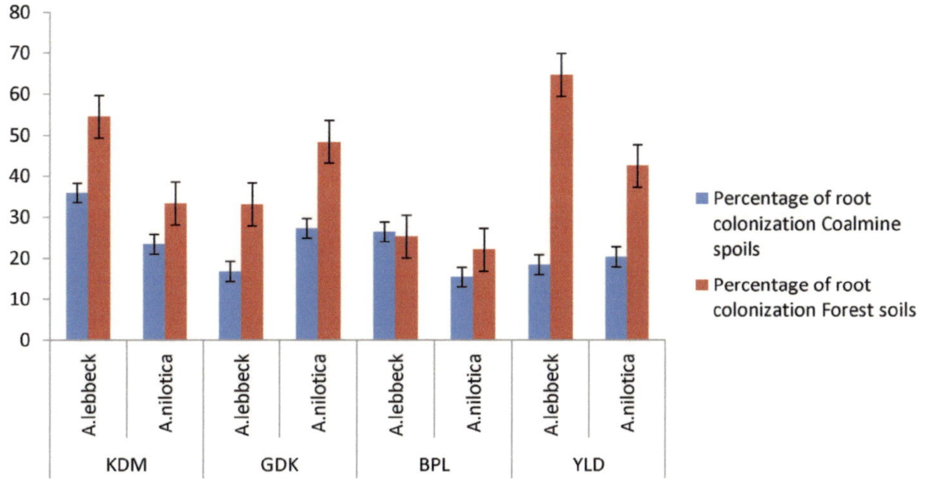

Figure 2.
Percentage of root colonization in two different soil types. KDM-Kothagudem, GDK- Godavarikhani, BPL-Bhopallaly, YLD-Yellandhu.

S. No	Treatments	Root colonization (%)	Height of plant (cm)	Biomass (g)		Phosphorus (mg/g)	
				Fresh wt.	Dry wt.	Shoot	Root
Albizia lebbeck							
1	*Glomus fasciculatum (Rhizophagus fasciculatum)*	72.6	112.1	125.2	108.8	0.36	0.12
2	*Glomus aggregatum (Rhizophagus aggregatus)*	76.4	120.2	142.6	116.3	0.42	0.38
3	*Gigaspora gigantea*	50.2	83.2	111.1	86.2	0.36	0.16
4	*Acaulospora foveata*	52.4	96.6	128.3	110.6	0.26	0.18
5	*Sclerocystis sp.*	39.7	86.5	98.2	65.3	0.14	0.10
6	*Control*	—	76.0	90.0	62.4	0.32	0.15
Acacia nilotica							
1	*Glomus fasciculatum (Rhizophagus fasciculatum)*	70.0	160.1	168.1	141.9	0.46	0.34
2	*Glomus aggregatum (Rhizophagus aggregatus)*	62.4	148.9	152.6	133.6	0.32	0.27
3	*Gigaspora gigantea*	69.6	154.2	156.1	139.2	0.41	0.29
4	*Acaulospora foveata*	50.4	120.2	136.2	99.3	0.25	0.16
5	*Sclerocystis sp.*	37.6	97.0	111.2	84.5	0.32	0.22
6	*Control*	—	68.8	72.8	59.1	0.22	0.12

Table 4.
Screening of Albizia lebbeck *and* Acacia nilotica *for efficient strains of AM fungi.*

less tough to soil disturbances than species from the suborder Gigasporineae, likely because of the qualities of either colonizing plant roots generally by hyphae or by spores.

Figure 2 shows the highest root colonization in the rhizosphere of *Albizia lebbeck* of Yellendu, while it was least in Bhopalpally soil. The percentage of root colonization was significantly varied with the type of soil. AM colonization was recorded moderately in the soil of Godavarikhani. No correlation could be observed between AM colonization and the spore population.

3. AM fungal inoculum development

AMF resting spores collected by stereo binocular microscope were surface sterilized with 200 ppm streptomycin for 15 min and washed in sterile distilled water for several times. The starter culture was prepared by soil funnel techniques [31]. A glass funnel was filled (3/4th) with autoclaved soil and sand (1:1) and the neck was loosely plugged with cotton wool. The funnel was kept over a conical flask (filled with sterile water). Spores were spread near the neck and covered with a thin layer of soil. Seeds of *Pennisetum glaucum* (surface-sterilized) were evenly sown and watered (sterile). After 10–15 days, the roots can be seen sprouting from the neck. Meanwhile, they get infected by AMF spores at the neck. After 25 days, roots were examined for root colonization.

When an adequate amount of growth was obtained by frequent sowing of seeds, this inoculum was transferred into small plastic pots filled with sand and soil (2:1) and mixed uniformly. Pots were transferred to the greenhouse and seeds of *Zea maize* and *Pennisetum glaucum* were sown. Pots were watered now and then with Hoagland nutrient solution without phosphorus and placed under uniform daylight. At the time of flowering of plants, the upper shoot system was cut off and fresh seeds were sown. After 2–3 months, the roots were mixed with soil and employed as inoculum for further experiments. **Figure 1** explains the isolation and screening of AM fungi for the selected host plants.

4. Screening of AM fungal species for efficiency

The effect of native AM fungi on the mycorrhizal intensity in terms of root colonization and spore number in rhizospheric soil of *Albizia lebbeck* and *Acacia nilotica* has been presented in **Table 4** respectively. In the comparative studies, all the *Glomus* species showed a significant difference in colonization.

Biomass of treated plants in the form of fresh weight recorded in *Albizia lebbeck* (*A.l*) and *Acacia nilotica* (*A.n*) is ranging from 90.0 to 142.6 g and 72.8 to 168.1 g, respectively. Likewise, root/shoot dry weight ranging from 62.4 to 116.3 g and 59.1 to 141.9 g, respectively, at the time of growth in the transplanted site. Minimum root/shoot growth was recorded in control plants. In comparison to control, all other treated plants showed the highest root/shoot growth. The maximum root/shoot growth of *Albizia lebbeck* (142.6 g) was recorded in *Glomus/Rhizophagus aggregatus* treatment. In *Acacia nilotica* (168.1), the highest growth was observed in *Glomus fasiculatum* and followed by *Gigaspora gigantea.*

In this study, all the five treatments gave the best results when compared with control (non inoculated tree species) *Glomus/Rhizophagus aggregatus* supports *Albizia lebbeck* showed the highest root colonization and helps the plants to uptake the nutrients such as root/shoot Phosphorus content (0.42/0.38 mg/g). In *Acacia nilotica* the highest shoot/root phosphorus content (0.46/0.34 mg/g) showed by treatment with *Glomus fasiculatum*.

Among all the five monoculture treatments *Glomus/Rhizophagus fasciculatum* and *Rhizophagus aggregatus* gave the best plant growth in all the parameters records plant height, biomass, and Phosphorus content. In this study percentage of AMF root colonization is directly proportional to the biomass and phosphorus content.

5. NPK fertilizers

The use of NPK fertilizers to enhance crop plant production include fertilization is specified to the soil and liquid forms of NPK that are sprayed on top of crop plants. In this time, plants are mostly fulfilled by giving solid fertilizers containing macronutrients, especially N inorganic continuously and without pains to restore the nutrients and absorb the essential elements with plants and causing the decrease of soil fertility [32, 33]. The use of extreme fertilizer is a waste of money and disturbs the stability of nutrients in the soil and increases environmental pollution [34, 35]. To improve the crop productivity and quality of outcome is required to be useful by reasonable fertilizer influences as a result that the proportion of nutrient absorption by plants is balanced and use of one type of fertilizer based on site-specific suggested doses [36]. The site-specific nutrient considers the potential of soils to give usual nutrients recovery [37]. To develop the nutrient status in the soil which administers the N-inorganic fertilizers in the required amounts of P, and K fertilizers are essential to increase crop production.

The compound fertilizers are containing the mineral elements, which need for the successful growth and development of plants. Mineral elements are necessary for optimal doses. Nitrogen, phosphorus, and potassium have enormous effects on plant growth and development. The deficiencies results indicate clear effects on the growth and yield of the crop plants. Nitrogen is a chlorophyll element, which promotes green color and vegetative growth of plants [44]. In agricultural systems, the most important crop nutrients are nitrogen (N), phosphorus (P), and potassium (K) [37].

Nitrogen fertilizing doses increase the protein levels and crop plant biomass, but the completion of N elements only without P and K will cause plants to simply drop, very sensitive to pest attacks disease, and reduced the quality of crop production [38]. Phosphorus nutrients in the soil absorbed by plants will be supported by P elements specified during fertilization [39]. Nutrient uptake of N, P, and K plants increases with an increasing dosage level of K fertilizer. Potassium is an important component involved in maintaining plant water conditions; it is responsible for regulating stomata opening and closing activities [40]. The multiple inorganic fertilizers added to plants can be either solid form or liquid. The spray of liquid fertilizer to the plants can play a role in improving the properties of the soil and supporting to enhance crop production [41, 42]. The application of liquid inorganic fertilizers is to make it an easy and efficient use of fertilizers by crop plants.

6. Effect of agrochemicals on interactions of Rhizobium and AM fungi on the growth of two forest trees

Different combination of agrochemicals (Captan, Sevin, TCP, Urea, 2, 4-D, DAP) along with Rhizobium and AM fungi were inoculated to test plants combinations are as follows Captan + *Glomus fasiculatum (G.f.)* **(A)**, Sevin + *G.f.* **(B)**, TCP + *G.f.* **(C)**, Urea + *G.f.* + *Rhizobium sp.* **(D)**, 2,4-D+ *G.f.* + *Rhizobium sp.* **(E)**, DAP + *G.f.* +Rhizobium sp. **(F)**, Control+ SS+ *Rhizobium sp.* **(G)**.

AM fungal infection was maximum in plants receiving treatment of E followed by D and F in descending order, while it was low in plants receiving treatment of A and B (**Table 5**). The spore population also increased in the presence of C, D, and F. Treatments C and D, promoted the plant growth. However, E promoted the maximum height followed by F and D plants, while it was least in B treated plants. Similarly, the treatment of E stimulated nodulation and biomass production. On the other hand, F and D influenced nodulation to an intermediate degree. The addition of tricalcium phosphate adversely affected growth-promoting activity. The degree of nodulation

Trestments	Infection(%)	No. of spores/ 100 g soil	Plant height (cm)	No. of nodules/ plant	Biomass		Phosphorus content (mg/plant)	
					Fresh wt.	Dry wt.	Shoot	Root
Albizia lebbeck								
Captan + G.f (A)	47.8 ± 1.28	83.0 ± 0.82	53.0 ± 0.82	—	22.5 ± 0.22	15.4 ± 0.16	0.15 ± 0.01	0.17 ± 0.01
Sevin+Gf.(B)	49.0 ± 0.54	104.0 ± 1.63	37.3 ± 1.25	—	20.8 ± 0.29	17.5 ± 0.25	0.28 ± 0.01	0.34 ± 0.02
TCP + G.f.(C)	42.5 ± 0.21	114.0 ± 0.82	54.3 ± 1.25	—	35.1 ± 0.53	26.3 ± 0.34	0.22 ± 0.01	0.35 ± 0.02
Urea + G.f. + Rhizobium sp.(D)	56.5 ± 1.25	118.0 ± 0.82	62.0 ± 1.63	46.0 ± 1.63	28.0 ± 1.30	18.7 ± 0.17	0.24 ± 0.01	0.34 ± 0.02
2,4-D+ G.f. + Rhizobium sp.(E)	58.5 ± 0.22	127.3 ± 1.70	66.0 ± 1.63	56.0 ± 1.63	38.3 ± 0.39	26.6 ± 0.42	0.36 ± 0.02	0.23 ± 0.02
DAP+G.f. + Rhizobium sp.(F)	56.4 ± 0.15	112.0 ± 1.63	64.0 ± 1.63	55.3 ± 2.49	37.3 ± 0.70	29.1 ± 0.29	0.14 ± 0.02	0.28 ± 0.01
Control+ SS + Rhizobium sp.(G)	46.5 ± 1.25	104.0 ± 1.63	48.0 ± 0.82	—	27.2 ± 0.82	16.4 ± 0.25	0.44 ± 0.22	0.47 ± 0.01
Acacia nilotica								
Captan + G.f (A)	55.2 ± 0.78	129.0 ± 0.82	52.9 ± 0.29	—	7.06 ± 0.01	3.92 ± 0.02	0.12 ± 0.01	0.15 ± 0.01
Sevin+Gf.(B)	64.5 ± 0.50	183.6 ± 2.05	64.4 ± 0.17	—	7.45 ± 0.02	4.14 ± 0.03	0.17 ± 0.01	0.14 ± 0.02
TCP + G.f.(C)	61.2 ± 0.49	165.6 ± 2.05	67.4 ± 0.87	—	8.75 ± 0.02	4.56 ± 0.01	0.21 ± 0.01	0.16 ± 0.02
Urea + G.f. + Rhizobium sp.(D)	68.1 ± 0.37	212.3 ± 1.70	67.6 ± 0.59	34.0 ± 1.63	9.16 ± 0.02	5.06 ± 0.01	0.22 ± 0.01	0.14 ± 0.01
2,4-D+ G.f. + Rhizobium sp.(E)	57.8 ± 1.25	154.3 ± 1.25	63.0 ± 0.62	37.6 ± 1.25	8.05 ± 0.03	6.26 ± 0.01	0.17 ± 0.01	0.13 ± 0.01
DAP+G.f. + Rhizobium sp.(F)	54.5 ± 0.68	122.6 ± 1.25	48.3 ± 1.03	27.6 ± 1.25	6.94 ± 0.01	3.72 ± 0.02	0.13 ± 0.01	0.91 ± 0.01
Control+ SS + Rhizobium sp.(G)	33.0 ± 0.62	109 ± 0.82	42.7 ± 0.37	—	5.42 ± 0.02	3.03 ± 0.02	0.91 ± 0.01	0.61 ± 0.01

Mean ± S.D. G.f. Glomus fasiculatum, and SS = Sterile soil.

Table 5.
Effect of agrochemicals on interactions of rhizobium and AM fungi on the growth of two forest trees.

decreased in the presence of captan and sevin which may also partly be due to the absence of *Rhizobium*. The biomass production varied with the agrochemicals tried. The maximum biomass production was recorded in the treatment E, while it was low in sevin treated plants. Similarly, the phosphorus content in shoot and root increased. The increase was more in root than in shoot. Treatments of F and C have adversely affected both the tree species. Maximum root infection was observed in *A. nilotica* plants receiving treatment D followed by B (**Table 5**). Root infection was least in E and F treatments. The stimulatory effect was comparatively more in treatment B than A. similar trend was observed in the spore population. The maximum spore population was recorded in plants receiving D treatment followed by B. It was low in F treated plants. Treatments A, C, and F adversely affected the AM infection and spore population.

The biomass production varied with the agrochemicals tried. The maximum AMF root colonization and biomass production was observed in the treatment E, while it was low in sevin treated plants. Similarly, Maximum plant height was recorded with fertilizers, while it was low in F treated plants. Nodulation increased in plants treated with E along with Rhizobium, while it was low in D and F. Treatments of A, B and C were responsible for inhibition of nodulation. On the other hand, the increased biomass production was recorded in plants treated with D and least in F. [43, 44] have recorded the adverse effect of some agrochemicals on AM colonization and growth and development of plants studied by them. Marginal change in physico-chemical characteristics of soils with the addition of different agrochemicals. The pH of the soils ranged between 7.0 and 8.1, and in C, E, and B soil was comparatively more alkaline. Maximum EC was recorded in D plants, while it was least in C plants. Organic matter was maximum in E plants, while it was low in B treated plants. Available phosphorus was also considerably increased in soils receiving different agrochemicals. Maximum available phosphorus was recorded in E treated soils followed by F and D treated plants. Available potassium was maximum in *G. fasciculatum* treated plants, followed by F plants and it was considerably low in D.

The root-based hyphal network in soils is the primary inoculum for seedlings that become established on natural grasslands. However, the inoculants colonized roots have some profound disadvantages since they may contain more than one mycorrhizal fungus and may also contain pathogenic organisms. Spores are possibly the best inoculants for laboratory experiments because the features diagnostic of individual species are present only in the spores developed primarily on extra metrical hyphae. Natural soil of crops and forests may contain varying numbers of spores of different AM fungal species. The dual culture using sterile soil with some kind of quality control is believed a practical approach to produce a high level of inoculants for commercial applications. A pot culture of *Glomus versiforme* on Sudan grass (*Sorghum vulgare*) can produce up to 107 spores per month over an extended period [45]. Spores from colonized soil near the colonized roots collected from field or pot cultures can be extracted using the traditional wet-sieve method. This approach and the later modified techniques are widely used in extracting spores from soils with modifications.

7. Applications of AM fungi

Arbuscular mycorrhizae show up as an exceptionally encouraging and monetarily reasonable device for the foundation of practical models of rural creation, because of their ability to expand the assimilation of fundamental supplements to plant development and increment their resistance to unfavorable ecological conditions, consequently keeping up soil quality and its gainful potential. Although enhancements in soil quality and plant nutritional status, for mycorrhizae application, was less investigated [46]. Various investigations have indicated that AM

Fungi can expand plant health and yield [47, 48]. These symbionts offer an eco-accommodating natural sound substitute to compound composts and pesticides for managing both plant quality and quantity in farming, cultivation, and ranger service. AM Fungi is currently viewed as the base of sustainable farming; so, there is a need to speed up its management in rural establishment frameworks [49].

The development of AM fungal hyphae is promoted by root exudates and is dependent on the arrangement of an appressorium increases the chance of hyphal entrance in the root framework. Dry weight and mycorrhizal dependence are the two most commonly utilized methods for assess AM fungal impact on plants [50]. Fungal impacts on plant physiology, for example, mineral nutrition especially phosphorus, plant execution, and plant assurance are significant segments in surveying contagious productivity.

AM fungi may similarly have connections among plant development advancing rhizosphere (PGPR) life forms. The impact of AM immunization may shift since numerous elements can impact the event of AM fungi [12].

8. Conclusion

The plant root infection and spore population were good in the forest soils and they were less in the overburden coal mine spoils. AMF exhibit different distribution patterns between these two soil types, where *Glomus* is dominant among all the species, *Scutellospora* and *Aculospora* were least in population. The high Mycorrhizal dependency value suggests that mycorrhizal inoculation would be useful in producing vigorous seedlings. In the nursery, which establish better and withstand some amount of drought and pathogenic infection. Seedlings inoculated with the indigenous AMF monoculture showed the highest biomass and phosphorus content when compared to non-mycorrhizal (controls), those plants grew very poorly. Within the AM fungi selection of perfect efficient indigenous mycorrhiza inoculations are needed for revegetation of disturbed sites. By the efficient AMF inoculation, the agroforestry tree species showed the best results in the form of increasing biomass and phosphorus uptake.

AM fungi and Plant Growth Promoting Rhizobacteria (PGPR) are significant parts in forest development and helps to increase biomass production [51]. There is a need of long term field studies to screen the efficent AM fungi in the revegetation sites and synergistic effects on indigenous microflora on tree growth.

Author details

Dayakar Govindu, Anusha Duvva and Srinivas Podeti*
Department of Biotechnology, Kakatiya University, Warangal, Telangana State, India

*Address all correspondence to: srinivas7586@gmail.com

References

[1] Zou, Y. N., Srivastava, A. K., Wu, Q. S. Glomalin: a potential soil conditioner for perennial fruits. *Int. J. Agric. Biol.* (2016). 18, 293-297. doi: 10.17957/IJAB/15.0085.

[2] Rodriguez A., Sanders I. R. (2015). The role of community and population ecology in applying mycorrhizal fungi for improved food security. ISME J. 9, 1053-1061. 10.1038/ismej.2014.207.

[3] Graham, J. H., Eissenstat, D. M. and Drouillard, D. L. 1991. On the relationship between a plant's mycorrhizal dependency and rate of vesicular-arbuscular colonization. Funct. Ecol. 5: 773-779

[4] Parniske M (2008) Arbuscular mycorrhiza: the mother of plant root endosymbioses. Nat Rev Microbiol 6: 763-775

[5] Olsson PA, Thingstrub I, Jakobsen I, Baath E (1999) Estimation of the biomass of arbuscular mycorrhizal fungi in a linseed field. Soil Biol Biochem 31:1879-1887

[6] Ortas I, Kaya Z, Çakmak I (2001) Influence of VA-mycorrhiza inoculation on growth of maize and green pepper plants in phosphorus and zinc deficient soils. In: Horst WJ, Schenk MK, Burkert A et al (eds.) Plant nutrition: Food security and sustainability of agro-ecosystems through basic and applied research. Kluwer Academic Publishers, Dordrecht, pp 632-633

[7] Ortas I (2003) Effect of selected mycorrhizal inoculation on phosphorus sustainability in sterile and non-sterile soils in the Harran plain in South Anatolia. J Plant Nutr 26:1-17. doi:10.1081/pln-120016494

[8] Aranda E, Scervino JM, Godoy P et al (2013) Role of arbuscular mycorrhizal fungus Rhizophagus Custos in the dissipation of PAHs under root-organ culture conditions. Environ Pollut 181:182-189. doi:10.1016/j.envpol.2013.06.034.

[9] Bowles TM, Barrios-Masias FH, Carlisle EA et al (2016) Effects of arbuscular mycorrhizae on tomato yield, nutrient uptake, water relations, and soil carbon dynamics under deficit irrigation in field conditions. Sci Total Environ 566:1223-1234. doi:10.1016/j.scitotenv.2016.05.178.

[10] Mehdi, Z., S.R. Nahid., H.A. Alikhani and A. Nasser. Responses of lentil to co-inoculation with phosphate-solubilizing rhizobial strains and arbuscular mycorrhizal fungi. J. Plant Nutr., (2006) 29: 1509-1522.

[11] Plenchette, C., Fortin, J.A. & Furlan, V. Growth responses of several plant species to mycorrhizae in a soil of moderate P-fertility. *Plant Soil* 70, 199-209 (1983).

[12] Abbott L.K, Robson. A.D (1991). Factors influencing the occurrence of vesicular arbuscular mycorrhizas agriculture ecosystems and environment 35,121-150.

[13] Wellings, N. P., & Thompson, J. P. (1993). EFFECTS OF VAM AND P FERTILIZER RATE ON. In *Proceedings of Second Asian Conference on Mycorrhiza: Bogor, Indonesia, 11-15 March 1991* (No. 42, p. 143).

[14] Sharma, M.P., Atimanav Guar, N.P. Bhatia, Alok Adholeya (1995). Response of Acacia nilotica to indigineous VAM fungal inoculation and single superphosphate fertilization in degraded waste soils. In mycorrhizae biofertilizers for the future. (eds). Proceedings of the third national conference on Mycorrhiza, pp. 534-540.

[15] Liu, D.; Liu, Y.; Fang, S.; Tian, Y. Tress species composition influenced

microbial diversity and nitrogen availability in rhizosphere soil. Plant Soil Environ. 2015, 10, 438-443

[16] Fernández, N.; Fontenla, S.; Messuti, M.I. Co-occurrence of arbuscular mycorrhizas and dark septate endophytes in pteridophytes from a Valdivian Temperate Rainforest in Patagonia, Argentina. In Mycorrhiza: Occurrence in Natural and Restored Environments; Pagano, M., Ed.; Nova Science Publishers: New York, NY, USA, 2011; pp. 99-126

[17] Smith, S.E.; Read, D.J. Mycorrhizal Symbiosis, 3rd ed.; Academic Press: New York, NY, USA, 2008; pp. 13-41.

[18] Cardoso, E. J. B. N., Vasconcellos, R. L. F., Bini, D., Miyauchi, M. Y. H., Santos, C. A. D., Alves, P. R. L., ... & Nogueira, M. A. (2013). Soil health: looking for suitable indicators. What should be considered to assess the effects of use and management on soil health?. *Scientia Agricola*, *70*(4), 274-289.

[19] Wicaksono, W.A.; Sansom, C.E.; Jones, E.E.; Perry, N.B.; Monk, J.; Ridgway, H.J. Arbuscular mycorrhizal fungi associated with *Leptospermum scoparium* (manuka): E ¯ ffects on plant growth and essential oil content. Symbiosis 2018, 75, 39-50.

[20] Zhang, T.; Hu, Y.J.; Zhang, K.; Tian, C.Y.; Guo, J.X. Arbuscular mycorrhizal fungi improve plant growth of *Ricinus communis* by altering photosynthetic properties and increasing pigments under drought and salt stress. Ind. Crop. Prod. 2018, 117, 13-19.

[21] Li, J.; Sun, Y.; Jiang, X.; Chen, B.; Zhang, X. Arbuscular mycorrhizal fungi alleviate arsenic toxicity to *Medicago sativa* by influencing arsenic speciation and partitioning. Ecotoxicol. Environ. Saf. 2018, 157, 235-243.

[22] Leifheit, E.F.; Veresoglou, S.D.; Lehmann, A.; Morris, E.K.; Rillig, M.C. Multiple factors fluence the role of arbuscular mycorrhizal fungi in soil aggregation-a meta-analysis. Plant Soil 2014, 374, 523-537.

[23] Van der Heijden, M.G.A.; Klironomos, J.N.; Ursic, M.; Moutoglis, P.; Strietwolf Engel, R.; Boller, T.; Wiemken, A.; Sanders, I.R. Mycorrhizal fungal diversity determines the plant diversity, ecosystem variability and productivity. Nature 1998, 398, 39-72.

[24] Bever, J.D.; Schultz, P.A.; Pringle, A.; Morton, H.B. Arbuscular mycorrhizal fungi: More diverse than meets the eye, and the ecological tale of why. Bioscience 2001, 51, 923-932.

[25] Jiang, J.; Moore, J.A.M.; Priyadarshi, A.; Classen, A.T. Plant-mycorrhizal interactions mediate plant community coexistence by altering resource demand. Ecology 2017, 98, 187-197.

[26] Jamiołkowska, A.; Ksi ˛e˙zniak, A.; Gał ˛azka, A.; Hetman, B.; Kopacki, M.; Skwary-Bednarz, B. Impact of abiotic factors on development of the community of arbuscular mycorrhizal fungi in the soil: A Review. Int. Agrophys. 2018, 32, 133-140.

[27] Chen, Y. L., Liu, R. J., Bi, Y. L., & Feng, G. (2014). Use of mycorrhizal fungi for forest plantations and minesite rehabilitation. In *Mycorrhizal Fungi: Use in Sustainable Agriculture and Land Restoration* (pp. 325-355). Springer, Berlin, Heidelberg.

[28] Wang, J., Wang, G. G., Zhang, B., Yuan, Z., Fu, Z., Yuan, Y., ... & Zhang, J. (2019). Arbuscular mycorrhizal fungi associated with tree species in a planted forest of Eastern China. *Forests*, *10*(5), 424.

[29] Zhao, Z.W.; Wang, G.H.; Yang, L. Biodiversity of arbuscular mycorrhizal fungi in a tropical rainforest of Xishuangbanna, southwest China. Fungal Divers. 2003, 13, 233-242.

[30] Hart, MM; Reader, RJ. Do arbuscular mycorrhizal fungi recover from soil disturbance differently? Tropical Ecology, 2004, 45, 97-111.

[31] T.H.Nicolson (1996).Vesicular-arbuscular Mycorrhiza -a plant symbiosis sci prog.,0xf,55,561-581

[32] S. L. Tisdale, Nelson and J.D. Beaton, *Soil Fertility and Fertilizers* (Machimilan Publising Co, New York, 1985),pp. 143-144.

[33] Bastari, *Penerapan Anjuran Teknologi Untuk Meningkatkan Efisiensi Penggunaan Pupuk* (Pusat Penelitian Tanah dan Agroklimat, Bogor, 1996), pp. 7-36.

[34] S. Adiningsih, J. S. Moersidi, M. Sudjadi and A.M. Fagi, "Evaluasi Keperluan Fosfat pada Lahan Sawah Intensifikasi di Jawa," in *Prosiding Lokakarya Nasional Efisiensi Penggunaan Pupuk* (Pusat Penelitian Tanah dan Agroklimat, Bogor, 1989), pp. 67-68.

[35] S. Rochayati, Muljadi and J.S. Sri Adiningsih, "Penelitian Efisiensi Penggunaan Pupuk di Lahan Sawah," in *Prosiding Lokakarya Nasional Efisiensi Penggunaan Pupuk* (Pusat Penelitian Tanah dan Agroklimat, Bogor, 1991), pp. 107-143.

[36] Dirjentan, "Program dan kebijakan pemerintah dalam pengembangan agribisnis jagung," in *Prosiding Seminar dan Lokakarya Nasional* (Pusat Penelitian dan Pengembangan Tanaman Pangan, Bogor, 2005), pp. 1-10.

[37] A. Dobermann and T. Fairthurts, *Rice nutrient disorders and nutrient management* (Internasional Rice Research Institute (IRRI), Los Banos, 2000), pp. 192.

[38] A. Rauf, B. M. Shepard and M. W. Johnson, Int. J. Pest. Manage 46, 257-266 (2000).

[39] Fi'liyah, Nurjaya and Syekhfani, Jurnal Tanah dan Sumberdaya Lahan 3(2), 329-337 (2016)

[40] Jones, J. B. Wolf and F. L. A. Mills, *Plant Analysis Handbook* (Micro-Macro Pub. Inc., USA, 1991), pp.213.

[41] Indrakusuma, *Proposal Pupuk Organik Cair Supra Alam Lestari* (PT Surya Pratama Alam, Yogyakarta, 2000),pp. 67.

[42] M. Sutedjo, *Pupuk dan Cara Pemupukan* (Rineka Cipta, Jakarta, 1999), pp. 45.

[43] Anasuya, D. J. Soil Biol. Ecol., 1996, 16, 35-39.

[44] .Rachel, E.K., Shailaja, K.M., Reddy, S.R. and Reddy, S.M. J.(1996). Effect of some agrochemicals of VAM infection and growth of sunflower. *Indian bot. Soc.,* 1996, 75, 179-181.

[45] B.A Daniels,A.Menge.(1981). Evaluation of the commercial potential of the vesicular arbuscular mycorrhizal fungus, Glomus epigaeus. New phytologist. *87*(2), 345-354.

[46] Da Silva Folli-Pereira M., Garlet J., Bertolazi A.A. Arbuscular Mycorrhizal Fungi and Their Potential Applications for Sustainable Agriculture. In: Yadav A., Mishra S., Kour D., Yadav N., Kumar A. (eds) Agriculturally Important Fungi for Sustainable Agriculture. Fungal Biology. Springer, Cham(2020)

[47] Mäder P., Edenhofer S., Boller T., Wiemken A., Niggli U.. Arbuscular mycorrhizae in a long-term field trial comparing low-input (organic, biological) and high-input (conventional) farming systems in a crop rotation. *Biol. Fertil. Soils* (2000) 31 150-156. 10.1007/s003740050638

[48] Hijri M "Analysis of a large dataset of mycorrhiza inoculation field trials

on potato shows highly significant increases in yield". Mycorrhiza. 2016 Apr; 26(3):209-14.

[49] Lone, Rafiq, et al. "Arbuscular mycorrhizal fungi for sustainable agriculture." *Probiotics and Plant Health*. Springer, Singapore, 2017. 553-577.

[50] Giovannetti, Manuela, et al. "Analysis of factors involved in fungal recognition responses to host-derived signals by arbuscular mycorrhizal fungi." *New Phytologist* 133.1 (1996): 65-71.

[51] Haselwandter, Kurt, and Glynn D. Bowen. "Mycorrhizal relations in trees for agroforestry and land rehabilitation." *Forest Ecology and Management* 81.1-3 (1996): 1-17.

Chapter 9

Influence of Endomycorrhizal Fungi on the Growth of Tropical Plant Species

Juan Francisco Aguirre-Medina, Jorge Cadena-Iñigue and Juan Francisco Aguirre-Caden

Abstract

In Southeast Mexico, deforestation in tropical forests is considered an environmental concern. Deforestation favours the growth of plants that compete with species of interest, which generally present slow growth. In order to promote greater growth in less time of the forest species used in reforestation and two crops of regional interest, the effect of including *Rhizophagus intraradices* in the seed or the root system of seedlings were investigated to evaluate its influence on growth and allocation of dry matter in the aerial and radical part, in addition to the concentration of N, P and Ca. Also, two additional collections with morphological characteristics of *Glomus* sps were included in *T. donnell-smithii*. The results indicate that biofertilisation with *R. intraradices* induces increased growth and is differential in stem and root dry matter allocation over time, in addition to increasing nutrient content in plant tissue.

Keywords: tropical spices, endomycorrhiza, root and shoot growth, range relative growth, nitrogen and phosphate

1. Introduction

Forest ecosystems are affected by various environmental and anthropogenic factors such as drought and the establishment of annual crops; these actions have caused their degradation. The above situation has generated several reforestation programs in Mexico through the massive production of forest species in nurseries, however, when planted in the field, the survival of plants is less than 50%, due in part to the low quality of the plants produced in the nursery and the intermittent drought they face when taken to the field [1]. In such procedures, the strengthening of the radical development of the host plant through biofertilisation with microorganisms has not been considered. Under these conditions, it is possible that microorganisms help plants survive in adverse environmental conditions [2].

Some bacteria and fungi that live in the rhizosphere interact with species present in agroforestry systems and may or may not manifest themselves in some morphological or physiological attribute of anthropocentric interest of the host plant, especially in sustainable or low input production systems, but their effective incidence depends on the microorganism and environmental and soil conditions [3]. The interaction of the plant with endomycorrhizal fungi stimulates its growth [4], even in adverse environmental conditions [3, 5], such as drought [6], presence of nematodes [7] and also activate defence mechanisms against various pathogens [8, 9].

Endomycorrhizal fungi interact symbiotically with more than 80% of terrestrial plants [10]. It is the most common symbiosis on earth [11] and important part of the development, maintenance and stability of ecosystems and represents an important mechanism for plant and reproductive development [12]. They can be found in all terrestrial ecosystems and their universality implies vast taxonomic diversity [13]. Native populations of endomycorrhizal fungi have favoured the sustainability of agricultural systems, while introduced ones may not be adapted to the environment and may have ecological specificity [14] or host preference [15]. In the rhizosphere, in addition to microorganisms, there are abundant root exudates that have a selective effect on soil microbiota [16].

When endomycorrhizal fungi are introduced in the seed or in the soil to colonise the root, the establishment of symbiosis is facilitated and the benefits are expressed in early stages [17], as an increase in growth and in the case of nursery plants, a decrease of time to be taken to the field. Symbiosis improves the supply, availability and physical access of nutrients to the plant [18].

Currently, endomycorrhizal fungi are considered essential organisms for the sustainable management of agriculture. In general, when new species are introduced to a region and adapt quickly to the new environment, the possibility that the species has the capacity in its root system to establish symbiosis with the microorganisms is considered. They are capable of linking to the development of the subway community [19].

Radical colonisation by endomycorrhizal fungi initiates with the exchange of carbon from the host plant to the fungus, and in turn, establishes by exploring the soil through mycelium the transport of nutrients to the root [20], such as phosphorus [2, 21, 22], water [20] and other nutrients to the plant. If Phosphorus (P) is not available for the initial development of the plant, it becomes limiting [23] and being a low mobility ion, hyphae can be the bridge for phosphorus supplementation [20] and by supplying it, growth is significantly influenced [2]. In addition, it improves the physical state of the soil by producing glomalin [20], to form more stable soil aggregates [24].

The beneficial effect of the application of endomycorrhizal fungi has been demonstrated in the increase of biomass in perennial crops such as *Leucaena leucocephala* (Lam.) De Wit, [25, 26], *Theobroma cacao* L. [27], *Coffea arabica* L. [28], *Jatropha curcas* L. [29], *C. canephora* (Pierre) ex Froehner [30], *Tabebuia donnell-smithii* (Rose) Miranda [31], *Cedrela odorata* L. [32] and *Elaeis guineensis* Jacq [33].

2. Influence of endomycorrhizal fungi on plant growth

Understanding growth, as the irreversible increase in the size of an individual almost always associated with an increase in its complexity, helps us to identify the effect of endomycorrhizal fungi on symbiosis with the host plant. The analysis of plant growth represents the first step in the analysis of primary production [34] in its different components, which are those that regulate the final production. In this way, the yield understood as the phenotypic expression of anthropocentric interest, is the final result of the physiological processes that are reflected in the plant's morphology [35].

The assignment of dry matter to the different structures of the plant, such as the root system, the stem and the foliar lamina are modified when endomycorrhizal fungi are included, either in soil or in substrates with the addition of other components, such as bovine manure, from the poultry industry, or agro-industrial wastes, such as sugarcane bagasse, coffee husk, cocoa shell, among others. This symbiotic association between fungus and plant generates the enlargement of the root system and acts as a root complement [36]. Endomycorrhizal fungi, together with the rest of microorganisms, are fundamental in the cycle of nutrients, even more, when considering that the availability of nutrients is heterogeneous in soils.

Mycelium is the means of transporting nutrients and water to the plant and is elemental in soil exploration. Especially in conditions of exploitation of monocultures that generate through time, the "depletion zone" of nutrients near its root system. Also, the tillage exercised in these production systems breaks the mycelium of the fungus and diminishes the beneficial effect it has on the structure of the soil, affecting the diversity and productivity of the communities [37, 38]. In addition, the applications of agrochemicals adversely affect the diversity and abundance of endomycorrhizal fungi in agroecosystems [39], causing the decrease or loss in the functioning of the plant community [40, 41].

In these conditions, the mycelium of the fungus can explore spaces in the soil where roots do not reach, that is, explore areas beyond the known "depletion zone" of the root and increase the absorption surface by exploring greater volume of soil compared to non-mycorrhised roots. This fact is more relevant when considering that they have the capacity to transport nutrients that are not very mobile, such as phosphorus [42].

The preference of endomycorrhizal fungi to transport phosphorus has been documented [21, 43] and this nutrient is fundamental for plant growth, even more so when considering that it has low availability in tropical soils [44]. Its supply by the fungus favours plant growth, but the symbiosis can be reduced or inhibited if the P level in the soil is high and the plant root can absorb it by itself [45]. [46] On the other hand, they cite that the efficiency in P absorption by the mycorrhised roots is mainly due to an acceleration of the dissociation of insoluble phosphate and it is extracted by the mycorrhizae according to the needs of the host plant. In general, it has been established that mycorrhised plants favour the absorption and transport of P, Zn, Ca, S, Cu, and Mg and their effects are more noticeable in low fertility soils [47].

When endomycorrhizal fungi are introduced into plants, the response can be diverse, influenced either by plant metabolism or by root architecture [48] and the interaction is considered non-specific, because any species of endomycorrhizal can colonise a plant [49], however, in different crops of the same species, the induction of growth is differential, according to the endomycorrhizal fungus introduced [31] or according to the tillage. In some forest species the root volume increases [50], in others, it decreases [51], or it is also expressed in an increase in the thickness of the stem needed to be taken to the field in less time compared to non-biofertilised plants. Also, it decreases mortality after transplantation [17] and improves its survival capacity in adverse conditions [52].

The growth of biofertilised plants with endomycorrhizal fungi presents changes in their aerial and root structure since the beginning of their evaluation. Generally, the dry weight of the aerial part of the plants is greater than the dry weight of the root system (**Figure 1**).

The root system development of biofertilised plants shows little difference in growth in *C. odorata* at 28 and 56 days after sowing. In contrast, the growth of *C. canephora*, increases and in *C. arabica*, its growth decreased at 28 days. The decrease in the growth of the root system of the biofertilised plants with the endomycorrhizal fungus in comparison to the control could reflect the initial limitation in the interaction of both organisms due to the availability of carbohydrates. In this initial stage, energy is required to support the mechanisms of plant/fungi recognition that will lead to the establishment of symbiosis [53], and in the

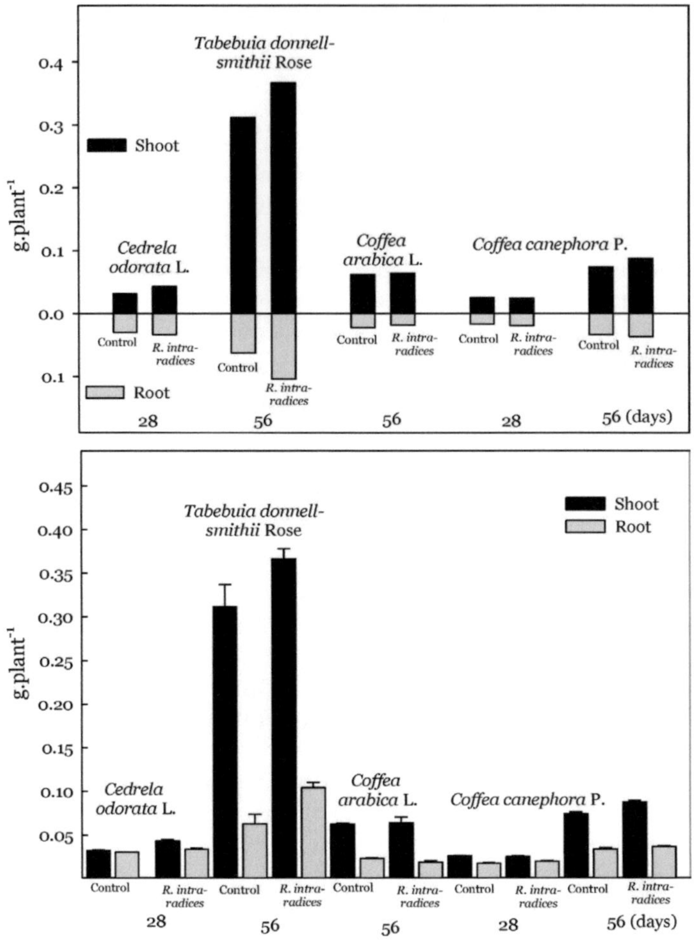

Figure 1.
Shoot and root growth of different tropical plant species with and without Rhizophagus intraradices *in the greenhouse. Values are the average of four replications by sampling and treatment. The vertical bars indicate the standard error of the mean.*

following stages, it is likely that the hyphae of the fungus will replace the root hairs due to the increase of biomass in the aerial part of the plant. Otherwise, it would be expected that the treatments with higher root system would have higher absorption of nutrients from the soil. The above behaviour suggests greater transport of photosynthates to the aerial part.

In *T. donnell-smithii,* the root system of the biofertilised treatment is increased. The values found at 140 das (**Figure 2**) indicate an increase in aerial and root biomass in most of the species evaluated with the exception of *T. cacao*. In other results with *T. cacao*, [54] the effect of the increase in biomass with the same species was found between 84 and 112 days. On the other hand, [55], they report increases in biomass of *T. cacao* with the inoculation of endomycorrhizal fungi of the *Scutellospora* and *Glomus* in genera of *T. cacao* plants with evident differences with the control from 120 days after sowing.

In *C. arabica*, the 140 das is presented with a slight increase in the radical system of the biofertilised plants in comparison to the control. In the same species [28] in evaluation at 60 and 90 days, there is a decrease in the biomass in the root system with the biofertilisation of *R. intraradices* at 60 and 90 days and a similar biomass in the root system as the control at 180 days. In contrast, with *C. canephora, T. donnell-smithii*, and *C. odorata*, the increase in aerial and radical biomass was remarkable.

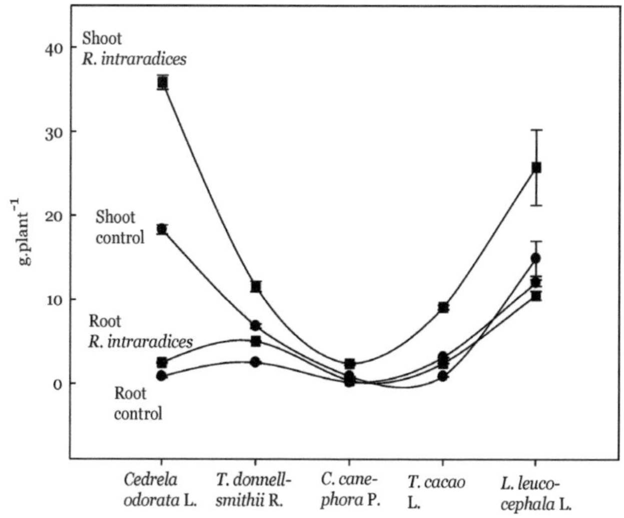

Figure 2.
Shoot and root growth of different tropical plant species with and without Rhizophagus intraradices *in the greenhouse. Values are the average of four replications by sampling. The vertical bars indicate the standard error of the mean.*

In *Leucaena leucocephala cv peru*, the radical system decreased with biofertilisation but the aerial biomass increased. It is probable that the plant responds better with other endomycorrhizal fungi.

The response of microorganisms in plant development reflects differential growth in time and between their organs. The greater or lesser allocation of biomass to some plant organs, seems to be influenced by the biofertilisation applied to its radical system and in almost all cases, after the ample development of an organ in a period of time, it tends to diminish in the following period, but with important increase in another plant organ.

The relative growth rate or the amount of dry matter produced per unit of time induced by the symbiosis with endomycorrhizal fungi presents the highest growth rate during the first 56 days after planting (days) (**Table 1**). In the period between 112 and 40 das, all species show a decrease in growth rate. This can be mainly related to the higher proportion of cells that do not divide in relation to those that do [56], whereas, in annual crops, the highest growth occurs around 30 days after biofertilisation with an increase in the mean relative growth rate [57] and leaf area [58].

In *T. cacao* L. and *C. canephora,* the relative growth rate presents a different effect between the control and *R. intraradices*. After the initial high growth with the endomycorrhizal fungus, it decreases during two months to continue again the increase of its growth. In the case of the control, the opposite happens, the initial increase in biomass is less, but it increases in the second and third sampling in *T. cacao* and the second in *C. canephora*. The above response is considered to be influenced by the demand for carbon sources [59].

In *T. donnell-smithii*, the response is different among fungal species. Some of them, such as *Glomus* sp. (Tea lemon), show a higher induction of initial growth and at the end of the evaluation, the growth was higher by *R. intraradices*. In this regard [60] cites differential response in growth [61] and, that plants have different responses to geographical isolations when the same species is inoculated.

On the other hand, in *C. odorata,* the growth rate was higher in the control on the evaluated dates. The physiological expression of plants seems to be related to their modular growth among their organs over time.

Plant species	Time (days after sowing)			
Cedrela odorata	28–56	56–84	84–112	112–140
Control	0.126*	0.042	0.054	0.021
R. intraradices	0.109	0.047	0.054	0.012
Tabebuia donnell-smithii	56–84	84–112	112–140	140–168
Control	0.065	0.022	0.019	0.012
R. intraradices	0.064	0.021	0.033	0.017
Glomus sp. Te Limón	0.073	0.022	0.023	0.014
Glomus sp. Vetiver	0.063	0.022	0.027	0.015
Teobroma cacao	30–60	60–90	90–120	120–150
Control	0.0070	0.0087	0.0072	0.0084
R. intraradices	0.0097	0.0076	0.0057	0.0085
Coffea canephora	28–56	56–84	84–112	112–140
Control	0.033	0.039	0.024	0.017
Rhizophagus	0.037	0.036	0.030	0.044

Values are the average of four replications by sampling and treatment.
**Dry weight ($g.g^{-1}.day^{-1}$).*

Table 1.
Relative growth rate (RGR) of the different tropical species biofertilised with R. intraradices *at the time of sowing under nursery conditions.*

3. Content of P and other nutrients in plant tissue

In all tropical forest species biofertilised with the endomycorrhizal fungus, a higher content of phosphorus was found in their plant tissue even though a large part of the available phosphorus in the soil is fixed in the andosol soils (**Table 2**). Many studies have shown that mycorrhised plants have benefits in their association with endomycorrhizal fungi under phosphorus-deficient conditions, especially in the acid soils of the tropics [62–64].

P is captured by the external mycelium and subsequently transported through the hyphae or intraradical structures in the form of polyphosphate granules and finally the process of transfer by the bush to the host cells [65].

In *C. arabica* plants at 56 days after planting, P and Ca values are very similar and concentrations increase to 140 days in the plant biofertilised with *R. intraradices*. With other species of endomycorrhizal fungi such as *Glomus clarum* and *Gigaspora margarita* in coffee seedlings, the growth and absorption of phosphorus by the plant was favoured, as well as the increase in survival and field production [66].

In general, it has been established that mycorrhised plants favour the absorption and transport not only of P but also of Zn, Ca, S, Cu, and Mg. The mycorrhizae are more active in soils of low fertility, especially when there is a deficiency of phosphorus [47].

The N content in the tissue of the biofertilised plants at 56 das was higher in the control with *C. arabica*. At 140 days, it was higher in most species with the exception of *C. canephora* cocoa plants biofertilised with *R. intraradices* had higher nitrogen content during the whole evaluation. This fact demonstrates that the host plant's root system is an extension of the plants' absorption system and favours the capacity to transport nutrients, such as nitrogen [18].

Plant species	Time (days)*	Nutrient (%)	
		N	P
Cedrela odorata L.			
Control	140	2.50	0.11
R. intraradices	140	3.11	0.19
T. donnell-smithii R.			
Control	140 (shoot)	0.74	0.08
	140 (root)	0.76	0.07
R. intraradices	140 (shoot)	0.73	0.12
	140 (root)	0.94	0.08
T. cacao L.			
Control	28	1.95	0.27
	56	1.77	0.10
	140	1.70	0.10
R. intraradices	28	2.62	0.33
	56	1.99	0.21
	140	1.83	0.11
C. arabica L.			
Control	56	1.98	0.13
	140	2.62	0.14
R. intraradices	56	1.88	0.14
	140	2.84	0.19
C. canephora P.			
Control	140	3.75	0.07
R. intraradices	140	3.55	0.131

Values are the average of four replications by sampling and treatment.
**Days after sowing.*

Table 2.
N, P and Ca content of different tropical species biofertilised with Rhizophagus intraradices *under greenhouse conditions.*

In the case of *T. donnell-smithii,* there are important differences between stem and root, in general, the highest N content was found in the root system. This difference in symbiosis effectiveness seems to depend more on the interaction with a soil type and crop conditions than with a particular host [67].

The benefits of transport of other nutrients and water, in addition to phosphorus to the plant, by mycorrhiza have been reported by several authors [68, 69].

Nowadays, the knowledge of microorganisms and their interaction with the rhizosphere has demonstrated the importance of symbiosis in the soil-plant system.

4. Radical colonisation

Plants with *R. intraradices* showed the highest initial mycorrhizal colonisation (45%) compared to (19%) in the control. The values of the control fluctuated between 4% (*C. arabica*) and 28% (*T. donnell-smithii*), in contrast with *R. intraradices,* the lowest value was also with *C. arabica* (18%) and the highest of 63, 57, 61 and 53% in *T. donnell-smithii*, *C. canephora*, *L. leucocephala* and *C. odorata,* respectively.

The radical colonisation in the controls confirms the presence of other endomycorrhizal fungi in the soils used, as part of the regional microbiota, but with less capacity to stimulate growth. Even though it has been indicated that the symbiosis lacks taxonomic specificity [70], there is a certain functional compatibility with the host plant, the substrate and the introduced microorganisms.

On the other hand, in biofertilised species, the speed of colonisation of *R. intraradices* is expressed as it has happened also in other annual and perennial crops [71] and consequently the photosynthetic activity increases after colonisation [20]. This fact suggests that the increase in the development of the host plant may be due to a greater capacity to absorb nutrients [32]. By introducing endomycorrhizal fungi attached to the seed, they have a greater chance of colonisation when the radicle emerges, unlike the fungi present in the substrate that may not be in the same proximity to the root system. However, it is expected that the symbiosis differs in the levels of colonisation [20] due to the interaction of environmental and management factors [2]. There are combinations of microorganisms that work best in a given host plant [70].

At the end of the evaluation (140 days) the percentage of colonisation in the control was 44% and in the treatment with *R. intraradices*, the average was 57%. The degree of benefit of the symbiosis may not be related to the percentage of colonisation.

5. Conclusions

The biofertilisation of tropical plants with *R. intraradices* favours the plant growth by increasing dry weight and the assignment of dry matter of the morphological and physiological components and the induction of growth is differential in time with respect to the organs evaluated, in some cases, it promotes greater aerial and radical growth and in others, it decreases the root system.

There are different periods in the vegetative growth of plants. At the beginning of the 28 to 56 days, the nutritional benefits are expressed by means of the increase in growth followed by a period of diminution and to continue in the following ones with the increase in biomass accumulation.

The content of nutrients such as phosphorus was always higher in the biofertilised plants.

Author details

Juan Francisco Aguirre-Medina[1*], Jorge Cadena-Iñigue[2]
and Juan Francisco Aguirre-Caden[1]

1 Autonomous University of Chiapas, Tapachula, Chiapas, Mexico

2 Postgraduate College, Texcoco, Mexico State, Mexico

*Address all correspondence to: juanf56@prodigy.net.mx

References

[1] Secretaría de Medio Ambiente y Recursos Naturales (SEMARNAT). (2005). Indicadores básicos del desempeño ambiental de México: [Internet]. Proyecto PNUD-SEMARNAT. México, D. F. [Accessed: 2013-02-12] http://app1.semarnat.gob.mx/dgeia/indicadores04/index.htm

[2] Andrade SAL, Mazzafera P, Schivinato MA, Silveira APD. Arbuscular mycorrhizal association in coffee. The Journal of Agricultural Science. 2009; 147(2): 105-115. DOI: https://doi.org/10.1017/S0021859608008344

[3] Qiu M, Zhang H, Wang G, Liu Z. Effects of nitrogen on plant-microorganism interaction. EurAsia J BioSci. 2008; 2, 4: 34-42. http://ejobios.org/download/effects-of-nitrogen-on-plant-microorganism-interaction.pdf

[4] Artursson V, Finlay RD, Jansson JK. Interactions between arbuscular mycorrhizal fungi and bacteria and their potential for stimulating plant growth. Environmental Microbiology. 2006; 8(1):1-10, DOI:10.1111/j.1462-2920.2005.00942.x

[5] Doubková P, Vlasáková E, Sudová R. Arbuscular mycorrhizal symbiosis alleviates drought stress imposed on *Knautia arvensis* plants in serpentine soil. Plant Soil. 2013; 370; 149-161. DOI: 10.1007/s11104-013-1610-7

[6] Augé RM. Arbuscular mycorrhizae and soil/plant water relations. Canadian Journal of Soil Science, 2004; 84(4): 373-381, https://DOI.org/10.4141/S04-002

[7] Dong LQ, Zhang KQ. Microbial control of plant-parasitic nematodes: a five-party Interaction. Plant and Soil. 2006; 288(1): 31-45. DOI: 10.1007/s11104-006-9009-3

[8] Harrier LA, Watson CA. The potential role of arbuscular mycorrhizal (AM) fungi in the bioprotection of plants against soil-borne pathogens in organic and/or other sustainable farming systems. Pest Management Science. 2004; 60 (2): 149-157. DOI: 10.1002/ps.820

[9] Jaizme-Vega MC, Rodríguez-Romero AS. Integración de microorganismos benéficos (Hongosmicorrícicos y bacterias rizosféricas) en agrosistemas de las Islas Canarias. Agroecología, 2008; 3: 33-39. Retrieved from http://revistas.um.es/agroecologia/article/view/95491/91801

[10] Gianinazzi S, Gollotte A, Binet M, van Tuinen D, Redecker D, Wipf D. Agroecology: the key role of arbuscular mycorrhizas in ecosystem services. Mycorrhiza, 2010; **20:** 519-530. https://doi.org/10.1007/s00572-010-0333-3

[11] Wang B, Qiu YL. Phylogenetic distribution and evolution of mycorrhizas in land plants. Mycorrhiza. 2006;16(5):299-363. DOI:10.1007/s00572-005-0033-6

[12] Read D. Plants on the web. Nature, 1998; 396: 22-23. https://doi.org/10.1038/23822

[13] Jaizme-Vega MC. Las micorrizas. Una simbiosis de interés para la agricultura. Revista Vida Rural. 2009; 288: 40-52. Retrieved from http://www.mapama. gob.es/ministerio/pags/Biblioteca/Revistas/pdf_Vrural%2FVrural_2009_288_48_52.pdf

[14] Serralde OAM, Ramírez GMM. Análisis de poblaciones de micorrizas en maíz (*Zea mays*) cultivado en suelos ácidos bajo diferentes tratamientos agronómicos. Revista Corpoica, 2004; 5: 31-40. https://doi.org/10.21930/rcta.vol5_num1_art:22

[15] Daniell T, Husband JR, Fitter AH, Young JPW. Molecular diversity of arbuscular mycorrhizal fungi colonizing arable crops. FEMS Microbiology Ecology. 2001; 36(2-3): 203-209. DOI: 10.1111/j.1574-6941.2001.tb00841.x

[16] Offre P, Pivato B, Siblot S, Gamalero E, Corberand T, Lemanceau P, Mougel C. Identification of bacterial groups preferentially associated with mycorrhizal roots of *Medicago truncatula*. Applied and Environmental Microbiology, 2007; 73, 913-921. DOI: 10.1128/AEM.02042-06

[17] Sieverding E. Ecology of VAM fungi in tropical agrosystems. Agriculture, Ecosystems & Environment, 1989; 29(1-4): 369-390. DOI: 10.1016/0167-8809(90)90303-U

[18] Barea JM, Azcon R, Azcon-Aguilar C. Mycorrhizosphere interactions to improve plant fitness and soil quality. Antonie van Leeuwenhoek Int. J. General Molecular Microbiol. 2002; 81(1-4): 343-351, DOI:10.1023/A:1020588701325

[19] Sanon A, Andrianjaka ZN, Prin Y, Bally R, Thioulouse J, Comte G, Duponnois R. Rhizosphere microbiota interfere with plant-plant interactions. Plant Soil, 2009; 321: 259-278. DOI: 10.1007/s11104-009-0010-5

[20] Sylvia MD. Mycorrhizal symbioses. In: Sylvia MD, Fuhrmann JJ, Harte GP, Zuberer AD, editors. Principles and applications of soil microbiology. 2nd ed. New Jersey, USA. Pearson Prentice Hall. 2005. p. 263-282.

[21] Marschner H, Dell B. Nutrient uptake in mycorrhizal symbiosis. Plant Soil. 1994; 159; 89-102 https://doi.org/10.1007/BF00000098

[22] Richardson AE, Barea JM, Mc Neill AM, Prigent-Combaret C. Acquisition of phosphorus and nitrogen in the rhizosphere and plant growth promotion by microorganisms. Plant Soil. 2009; 321:305-339, DOI 10.1007/s11104-009-9895-2

[23] Tristão FSM, López ASA, Silveira APD. Fungos micorrízicos arbusculares na formação de mudas de cafeeiro, em substratos orgânicos comerciais. Bragantia, 2006; 65(4): 649-658. DOI: 10.1590/S0006-87052006000400016

[24] Rillig MC, Mummey DL. Mycorrhizas and soil structure. New Phytologist, 2006; 171(1): 41-53. DOI: 10.1111/j.1469-8137.2006.01750.x

[25] Ruiz-Torres GJ. La inoculación de *Leucaena leucocephala* con *Glomus intraradices* y su efecto en la asignación y calidad de la materia seca. [Tesis]. Facultad de Ciencias Agrícolas. Universidad Autónoma de Chiapas. 2005.

[26] Aguirre-Medina JF, Gálvez-López AL, Ibarra-Puón JC. Growth of *Leucaena leucocephala* (Lam.) de Wit biofertilized with arbuscular mycorrhizal fungi in the nursery. Revista Chapingo Serie Ciencias Forestales y del Ambiente. 2018; 24(1): 49-58. DOI: 10.5154/r.rchscfa.2017.07.043

[27] Aguirre-Medina Juan Francisco, Mendoza-López Alexander, Cadena-Iñiguez Jorge, Avendaño-Arrazate Carlos H. Efecto de la biofertilización en vivero del cacao (*Theobroma cacao* l) con *Azospirillum basilense* tarrand, krieg et Döbereiner y *Glomus intraradices* Schenk et Smith. Interciencia. 2007; 32 (8): 541-546. Recuperado de: https://www.redalyc.org/articulo.oa?id=33932808

[28] Aguirre-Medina JF, Moroyoqui-Ovilla DM, Mendoza-López A, Cadena-Iñiguez J, Avendaño-Arrazate CH, Aguirre-Cadena JF. Hongo endomicorrízico y bacteria fijadora de

nitrógeno inoculadas a *Coffea arabica* en vivero. Agronomía Mesoamericana. 2011; 22 (1): 71-80. https://www.scielo.sa.cr/pdf/am/v22n1/a09v22n1.pdf

[29] Díaz-Hernández BG, Aguirre-Medina JF, Díaz-Fuentes VH. Rendimiento de *Jatropha curcas* L. inoculada con micorriza y aplicación de composta de caña. Revista Mexicana de Ciencias Agrícolas. 2013; 4 (4):599-610. http://www.scielo.org.mx/pdf/remexca/v4n4/v4n4a9.pdf

[30] Ibarra-Puón, JC, Aguirre-Medina JF, Ley-De Coss A, Cadena-Iñiguez J, Zavala-MataA. Inoculación de Coffea canephora (Pierre) ex Froehner con Rhizophagus intraradices (Schenck et Sm.) Walker et Schuessler y Azospirillum brasilense Tarrand, Krieg et Döbereiner en vivero. Revista Chapingo Serie Horticultura, 2014; 20(2); 201-213. DOI: 10.5154/r.rchsh.2013.09.027

[31] Aguirre-Medina JF, Culebro-Cifuentes F, Cadena-Iñiguez J, Aguirre-Cadena JF. Crecimiento de *Tabebuia Donnell-Smithii* (Rose) Inoculada con Hongos Micorrizicos y *Azospirillum brasilense*. Agrociencia. 2014; 48 (3):331-345. http://www.scielo.org.mx/pdf/agro/v48n3/v48n3a8.pdf

[32] Aguirre-Medina JF, Mina-Briones F, Cadena-Iñiguez J, Dardón-Zunun JD, Hernández-Sedas DA. Crecimiento de *Cedrela odorata* L. biofertilizada con Rhizophagus intraradices y Azospirillum brasilense en vivero. Revista Chapingo Serie Ciencias Forestales y del Ambiente. 2014; 20(3): 177-186. DOI: 10.5154/r.rchscfa.2014.01.001

[33] Garza-Hernández JM, Marroquín-Agreda FJ, Lerma-Molina JN, Ley de-Coss A, Toledo-Toledo E, Martínez-Solís M, Villalobos-Villalobos V, Aguirre-Medina JF. Biofertilizante micorrízico y fertilizante mineral en el crecimiento de Elaeis guineensis Jacq en vivero. AGROproductividad. 2016; 9 (2): 26-32. http://www.revista-agroproductividad.org/index.php/agroproductividad/article/view/717/586

[34] Kvet J, Ondok JP, Necas J, Jarvis PG. 1971. Methods of growth analysis. *In*: Plant photosynthetic production. Manual of Methods. Sestak Z, Catsky J, Jarvis PG. editors. The Hague Netherlands. p. 343-384.

[35] Kohashi-Shibata J. Fisiología. In: Engleman EM, Editor. Contribuciones al conocimiento del fríjol *Phaseolus* en México: Chapingo, México Colegio de Postgraduados; 1979. p. 39-58.

[36] Colozzi Filho A, Cardoso Elke. Detecção de fungos micorrizicos arbusculares em raízes de cafeeiro e de crotolária cultivada na entrelinha. Pesquisa Agropecuária Brasileira. 2000; 35(10): 2033-2042. https://DOI.org/10.1590/S0100-204X2000001000015

[37] Hijri B, Sykorova Z, Oehl F, Ineichen K, Mäder P, Wiemkwm A, Redecker D. Communities of arbuscular mycorrhizal fungi in arable soils are not necessarily low in diversity. Molecular Ecology 2006; 15: 2277-2289. DOI: 10.1111/j.1365-294X.2006.02921.x

[38] Dandan Z, Zhiwei Z. Biodiversity of arbuscular mycorrhizal fungi in the hot-dry Valley of the Jinsha River, Southwest China. Applied Soil Ecology. 2007; 37: 118-128. https://doi.org/10.1016/j.apsoil.2007.06.003

[39] Douds DD, and P Millner. Biodiversity of arbuscular mycorrhizal fungi in agroecosystems. Agric. Ecosystems Environ. 1999; 74: 77-93. https://doi.org/10.1016/B978-0-444-50019-9.50008-X

[40] Hart MM, Reader RJ. Taxonomic basis for variation in the colonization strategy of arbuscular mycorrhizal fungi. New Phytologist. 2002; 153: 335-344 https://nph.

onlinelibrary.wiley.com/doi/pdf/10.1046/j.0028-646X.2001.00312.x

[41] Munkvold LR, Kjoller M, Vestberg S, Rosendahl S, Jakobsen. Hight functional diversity within species of arbuscular mycorrhizal fungi. New Phytol. 2004; 164: 357-364. https://doi.org/10.1111/j.1469-8137.2004.01169.x

[42] Sylvia D, Fuhrmann J, Hartel P, Zuberer D. Principles and applications of soil microbiology. New Jersey USA: Ed. Prentice Hall Inc; 1999. 550 p.

[43] Zhu Y, Smith AF, Smith SE. Phosphorus efficiencies and responses of barley (*Hordeum vulgare* L.) to arbuscular mycorrhizal fungi grown in highly calcareous soil. Mycorrhiza. 2003; 13: 93-100. https://doi.org/10.1007/s00572-002-0205-6

[44] Rippstein G, Amezquita E, Escobar E, Grollier C. Condiciones naturales de la sabana. In: Rippstein G, Escobar G, Mota F, editors. Agroecología y Biodiversidad de las sabanas en los llanos orientales de Colombia. Cali Colombia: International Center for Tropical Agriculture. 2001. p. 1-21. https://cgspace.cgiar.org/bitstream/handle/10568/55160/agroecologia_y_biodiversidad.pdf?sequence=1&isAllowed=y

[45] Blancof F, Salas E. Micorrizas en la agricultura: contexto mundial e investigación realizada en Costa Rica. Agronomía Costarricense. 1997; 21(1): 55-67. Consultado el 15 de Agosto 2020 http://listas.exa.unne.edu.ar/biologia/fisiologia.vegetal/Micorrizas%20y%20nutricion%20mineral.pdf

[46] Miyasaka S, Habte M. Miyasaka SC, Habte M, Friday JB, Johnson EV. Manual on Arbuscular Mycorrhizal Fungus Production and Inoculation Techniques. College of Tropical Agriculture and Human Resources SCM-5. 2003; p.1-4. https://scholarspace.manoa.hawaii.edu/bitstream/10125/12454/SCM-5.pdf

[47] Gerdemann JW. Vesicular-Arbuscular mycorrhiza and Plant growth. Annual Review of Phytopathology, 1968; 6(1): 397-418. DOI: 10.1146/annurev.py.06.090168.002145

[48] Dodd J. 1999. Micorrizas. Recent advances in understanding the role of arbuscular mycorrhizas in plant production. In: Siqueira JO et al. editors. Interrelação fertilidade, biologia do solo e nutrição de plantas: Lavras Sbs/Ufla; 1999. p. 687-703.

[49] Molina M, MaecHa L, Medina M. Importancia del manejo de hongos micorrizógenos en el establecimiento de árboles en sistemas silvopastoriles. Revista colombiana de ciencias pecuarias, 2005; 18(2): 162-175. https://www.researchgate.net/publication/262670903_Importance_of_mycorrhizae_

[50] Bowen GD, and AD Rovira. The Rhizosphere and its management to improve plant growth. Advances in agronomy. 1999; 66: 1-102. https://DOI.org/10.1016/S0065-2113(08)60425-3

[51] Mendoza López A. La biofertilización del cacao *Theobroma cacao* L. en etapa de vivero con *Azospirillum brasilense* Tarrand, Krieg *et* Dobereiner y *Glomus intraradices* Schenk70 et Smith. [Tesis]. Facultad de Ciencias Agrícolas Huehuetan Chiapas, México: Universidad Autónoma de Chiapas. 2003.

[52] Aguirre-Medina, J. F., Kohashi-Shibata, J., Trejo-López, C., Acosta Gallegos, J. A., & Cadena-Iñiguez J. (2005). Inoculación de Phaseolus vulgaris L. con tres microorganismos y su efecto en tolerancia a sequía. Agricultura Técnica en México, 31, 125-137. Retrieved from https://biblat.unam.mx/es/revista/agricultura-tecnica-en-mexico/9

[53] Bonfante-Fassolo P, and Perotto S. Plants and endomycorrhizal fungi.

The cellular and molecular basis of their interaction. In: Verma DP, Editor. Molecular signals in plant-microbe communications. Boca Raton, Fla, CRC press; 1992. p. 445-470.

[54] Mina Briones FO. Crecimiento de *Cedrela odorata* L. con diferentes microorganismos inoculados a la semilla en vivero. [Tesis]. Facultad de Ciencias Agrícolas, Huehuetán, Chiapas. México: Universidad Autónoma de Chiapas. 2013.

[55] Chulan HA, Martin K. The vesicular-arbuscular (VA) mycorrhiza and its effects on growth of vegetatively propagated *Theobroma cacao* L. Plant Soil. 1992; 144: 227-233. https://doi.org/10.1007/BF00012879

[56] Milthorpe FL, Moorby J. Introducción a la fisiología de los cultivos. Argentina: Ed. Hemisferio Sur; 1982. 259 p.

[57] Koucheki HK, Read DS. Vesicular-arbuscular mycorrhiza in natural vegetation systems. II. The relationship between infection and growth in *Festuca ovina* L. The New Phytologist. 1976; 77 (3): 655-666. https://doi.org/10.1111/j.1469-8137.1976.tb04658.x

[58] Allen MF, Moore TS Jr, Christensen M. Phytohomone change *in Bouteloua gracilis* infected by vesicular-arbuscular mycorrhizal. II. Altered levels of gibberellin-like substances and abscisic acid in the host plant. Canadian Journal of Botany. 1982; 60(4): 468-471, https://DOI.org/10.1139/b82-063

[59] Sprent JI, de Faria SM. Mechanisms of infection of plants by nitrogen fixing organisms. Plant Soil. 1998; 110: 157-165. https://doi.org/10.1007/BF02226795

[60] Carling DE, Brown MF, Brown RA. Colonization rates and growth responses to soybean plants infected by vesicular-arbuscular fungi. Canadian Journal of Botany. 1979; 57(17): 1769-1772, https://DOI.org/10.1139/b79-218

[61] Bethlenfalvay GS. Mycorrhizae in the agricultural plant-soil system. Symbiosis. 1992; 14: 413-425. https://dalspace.library.dal.ca/bitstream/handle/10222/77317/VOLUME%20 14-NUMBERS%201,2,3-1993-PAGE%20 413.pdf?sequence=1

[62] Gyaneshwar P, Naresh Kumar, Parekh G, LJ, Poole P. Role of soil microorganisms in improving P nutrition of plants. Plant and Soil, 2002; 245: 83-93. https://doi.org/10.1023/A:1020663916259

[63] Smith S, Anderson I, Smith F. Mycorrhizal associations and phosphorus acquisition: from cells to ecosystems. Annual Plants Review. 2015; 48: 409-440. https://doi.org/10.1002/9781118958841.ch14

[64] Aguirre-Medina JF, Kohashi-Shibata J. Dinámica de la colonización micorrizica y su efecto sobre los componentes del rendimiento y el contenido de fósforo en frijol común. Agricultura Técnica en México. 2002; 28 (1): 23-33. https://biblat.unam.mx/es/revista/agricultura-tecnica-en-mexico/13

[65] Mosse B. Advances in the study of VA micorrhiza. Ann. Rev. Phytopath. 1973; 11: 171-196. https://doi.org/10.1146/annurev.py.11.090173.001131

[66] Siqueira JO, Colozzi–Filho A, Saggin OJ, Guimaraes TG, Oliveira E. Crescimento de mudas e producao do cafeeiro sob influencia de fungos micorrízicos e superfosfato. Revista Brasileira Ciencia Solo. 1993; 17(1): 53-60. file:///C:/Users/JF%20Aguirre/Downloads/Siqueiraetalcrescimentodemudas RBCSv17n11993.pdf

[67] Azcón-G. de Aguilar C, Barea JM. Micorrizas. Investigación y Ciencia. 1980; 47: p. 8-16. https://www.investigacionyciencia.es/revistas/investigacion-y-ciencia/chaquete-90

[68] Rhodes LH, Gerdemann JW. Translocation of calcium and phosphate by external hiphae of vesicular-arbuscular mycorrhiza. Soil Sci. 1978; 126 (2): 125-126. https://journals.lww.com/soilsci/Abstract/1978/08000/TRANSLOCATION_OF_CALCIUM_AND_PHOSPHATE_BY_EXTERNAL.9.aspx

[69] Aguirre-Medina JF. Biofertilizantes microbianos: Experiencias agronómicas del programa nacional del INIFAP en México. Rosario Izapa, Chiapas México. Instituto Nacional de Investigaciones Forestales, Agrícolas y Pecuarias 2006. 201 p. Retrieved from http://biblioteca.inifap.gob.mx:8080/jspui/handle/123456789/3633

[70] Aguirre-Medina JF, Aguirre-Cadena JF, Cadena-Iñiguez J, Avendaño-Arrazate CH. Biofertilización en plantas de la selva húmeda tropical. México: Colegio de Postgraduados. 2012. 99 p.

[71] Cuenca G, Cáceres A, Oirdobro G, Hasmy Z, Urdaneta C. Las micorrizas arbusculares como alternativa para una agricultura sustentable en áreas tropicales. Interciencia 2007; 32(1): 23-29. http://www.interciencia.org/v32_01/23.pdf